THE SHAPING OF AMERICAN LIBERALISM

DAVID F. ERICSON

THE SHAPING OF AMERICAN LIBERALISM

THE DEBATES OVER RATIFICATION NULLIFICATION AND SLAVERY

THE UNIVERSITY OF CHICAGO PRESS
CHICAGO AND LONDON

DAVID F. ERICSON is assistant professor in the Department of Political Science at Wichita State University.

The University of Chicago Press, Chicago 60637
The University of Chicago Press, Ltd., London

Printed in the United States of America
02 01 00 99 98 97 96 95 94 93   1 2 3 4 5

ISBN (cloth): 0-226-21683-7
ISBN (paper): 0-226-21684-5

Library of Congress Cataloging-in-Publication Data

Ericson, David F., 1950-
The shaping of American liberalism: The debates over ratification, nullification, and slavery /
David F. Ericson.
p. cm.
Includes bibliographical references and index.
1. United States—Politics and government—1783–1865. 2. United States—Constitutional history. 3. Nullification. 4. Lincoln-Douglas debates, 1858. I. Title.
E302. 1.E74 1993
320.973—dc20   92–31976
CIP

∞ The paper used in this publication meets the minimum requirements of the American National Standard for Information Sciences—Permanence of Paper for Printed Library Materials, ANSI Z39.48–1984.

In memory of J. David Greenstone

Contents

Acknowledgments

I would like to thank Nathan Tarcov, Joseph Cropsey, and Liane Kosaski for their comments on earlier (and, in Nathan's case, almost final) versions of this manuscript. Part of the research for this book was funded by a postdoctoral fellowship from the John M. Olin Center for the Study of the History of Political Culture at the University of Chicago. I would like to thank Stephen Holmes for making that possible. I would also like to thank my editors at The University of Chicago Press: John Tryneski, Randolph Petilos, and Kathryn Kraynik. But, above all, this book would not have been possible—it would not have even been conceived!—without the influence of J. David Greenstone.

Careful readers of this book and of David's forthcoming book, *The Lincoln Persuasion: Politics and Synthesis in American Politics,* will notice many similarities as well as many differences between the two works. The time frame of their research and writing was almost coincident. We engaged in a very lively interchange of ideas over the ten years I was at the University of Chicago and various places in between. We never could agree on Webster and Lincoln, or on the general nature of American liberalism. His most important influences on me, however, concerned the manner, more than the substance, of our conversations. He taught me that the most important thing was not to agree but to care.

1

Republicanism as a Political Tradition

Louis Hartz's *The Liberal Tradition in America* set the tone for a generation of scholarship on American political thought.[1] Hartz argued that the history of American political thought was a projection of its original liberal principles. Of course, no thesis stands long without its revisionists. Starting with Bernard Bailyn's *The Ideological Origins of the American Revolution,* the republican revisionists have chipped away at Hartz's thesis.[2] They claim that American political thought, at least into the 1780s, was dominated not by liberal principles but by preliberal, republican principles. In the meantime, the liberal thesis has been reasserted by, among others, Joyce Appleby, though much more circumspectly than Hartz had initially asserted it.[3] These neoliberals are currently skirmishing the republican revisionists over the "souls" of different American political figures and parties, and both sides are alternately extending and restricting the reach of their historical arguments. An uneasy truce has developed in which each side admits the partial validity of the other's thesis. They agree that the history of American political thought has been the battleground for classical republican and modern liberal ideas. The debate now is over who scored victories when in this intellectual warfare.[4]

While I agree with the republican revisionists that American political thought has not been simply liberal, I also think that Hartz was right about the consensually liberal nature of American political thought. The resolution of this apparent paradox lies in considering four "framing" questions which have not been satisfactorily addressed in the secondary literature: (1) What does republicanism mean?; (2) What is its relation to liberalism?; (3) Is the republican thesis a consensus view (at least for some time period) and what does that mean?; and (4) What does it mean to call a particular historical actor a republican?

I will offer a specific definition of republicanism in the next chapter. This definition will attempt to synthesize the various ways in which other scholars have, more loosely, used the term. Undoubtedly, none of them would completely agree with my definition. The broader point, however, is that some, fairly specific definition is necessary to set the terms of debate in analyzing the history of American political thought, either in whole or part.[5]

The contested nature of the republican thesis magnifies this necessity. The arguments over its relation to the liberal thesis which it revised have, in fact, produced more sparks than the arguments over whether particular historical actors were liberals or republicans. Pocock's heated defense of his own *Machiavellian Moment* is only the most extreme example of a widespread tendency that threatens to transform a debate about history into a debate exclusively about historiography.[6]

As I see it, republicanism is related to liberalism as species to genus. Historically, there have been different species of liberalism or, in long hand, different ways of constructing a working public philosophy out of the rich universe of liberal ideas created by such early-modern philosophers as Thomas Hobbes and John Locke and henceforth available to political actors. If the liberal universe revolved around the effort to draw an ideological boundary between what is legitimately public and private, then such political actors as Thomas Jefferson and John Adams could collect a set of ideas from that universe which drew that boundary at significantly different places. I define republicanism as a species of liberalism which granted relatively more space to the public sphere in distinction to pluralism, a species of liberalism which granted relatively more space to the private sphere. I, thus, define republicanism and pluralism as two species of liberalism in much the same way other scholars refer to republicanism and liberalism as two genera of political thought.[7]

This redefinition is not a mere juggle on my part. What is at stake is whether the historiographical debate over the validity of the republican and liberal theses concerns the relative influence of two radically different universes of ideas on American political thought or of two different sets of ideas with common roots. Does it concern the relative influence of classical republicanism and modern liberalism or of two forms of the latter? Depending on how we define the terms of the debate, it assumes a very different character, and result. If the terms are classical republicanism and modern liberalism, the republican thesis is the clear loser. But if they are modern republicanism and pluralism, the republican thesis is the victor, at least for the late eighteenth and early nineteenth centuries. It is the caricature of what the American revolutionaries believed and of what liberals must believe that has distorted the historiographical debate. The most the republican revisionists have shown is that a few revolutionary actors spoke of creating a Christian Sparta in America, not that even they offered that as a serious political possibility; just as the most the neoliberals have shown is that many Americans thought the protection and promotion of private rights and interests was essential to the new political order, not that they considered that goal definitive of the new order.[8]

This confusion in the secondary literature is, of course, not groundless. There *was* a classical republicanism. Modern republicanism *does* bear some affinities to it. Classical and modern republicans share the same preference for a small-scale, participatory politics, and they share the same ambivalence toward commerce and factions. Yet, the fundamental beliefs of modern republicans are liberal ones, hence *not* classical ones. Modern republicans like Jefferson and Adams were committed to liberty and happiness as the primary political ends, not, for instance, to attaining a Ciceronian "partnership in Justice." Accordingly, there is in modern, as opposed to classical, republicanism a powerful impetus toward political democracy. It is these areas of divergence between modern and classical republicanism which both the republican revisionists and neoliberals overlook.[9]

These provisional conclusions raise the issue of consensus. This issue is complicated by the three levels on which consensus is possible within the (redefined) terms of the historiographical debate. On these three levels, my claims are that: (1) there was (is) a fundamental liberal consensus in America; (2) a more specific republican consensus existed in America during the late eighteenth and early nineteenth centuries, but by the time of the Civil War a consensus shift to pluralism had occurred; and (3) even within those second-level consensuses, there were significant divisions, as between federal (states-rights) and national republicans and moderate and extreme pluralists.[10]

By these claims, I do not mean that all Americans were liberals or, at one time, liberal republicans. The relevant sample(s) can only include Americans for whom evidence exists that they thought about politics in coherent enough terms to raise the question of the character of their political thought. Consensus, therefore, refers to those few Americans, who I will call statesmen. This term is intended to demarcate a middle category between political philosophers and ordinary politicians, for which James Madison would be the American exemplar. To treat the ideas of these "few" as representative of the "many" is comforting and, to a certain extent, doubtlessly valid. Ultimately, though, the ideas of the two groups are incommensurate.[11]

Consensus, furthermore, refers to only "most" of a few. A liberal consensus does not preclude nonliberals, nor does a republican consensus preclude pluralists (or vice versa). For example, Orestes Brownson's Christian socialism did not belie a broad-based liberal consensus among Jacksonian party leaders, nor did William Leggett's free-market pluralism belie a more specific republican consensus among them.[12]

Finally, consensus does not prevent divisive conflict. In chapters 3 through 8, I, indeed, will analyze three critical debates in American history as debates which were fueled by splits within American liberalism and among, first, republicans and, then, pluralists. It has often been said that the major debates leading up to the Civil War were debates over the meaning of American nationality. The republican thesis helps us see why.[13] One of the primary intellectual forces behind the ratification and nullification debates was the disproportion between American nationality and the traditional republican preference for a small-scale, participatory politics. Federal republicans expressed serious concerns about that disproportion; national republicans sought to assuage those concerns using the same political idiom. The common frame of reference itself stoked the fires of controversy. By the time the slavery issue took centerstage in the 1850s, the debate over American nationality had assumed a very different form. While there were still powerful echoes of earlier nationality debates, the long-running dispute between federal and national republicans had essentially been won by the latter. As a result, the slavery issue was largely debated in terms of how pluralistic American society had (or should) become. For some, like Abraham Lincoln, the continued existence of slavery even called into question the nation's basic commitment to liberal principles.[14]

For each of these debates—ratification, nullification, and slavery—I will closely analyze the "outstanding" representative of each side. My sample will include the Federal Farmer and Publius; John C. Calhoun and Daniel Webster; Stephen A. Douglas and Lincoln. These choices are hardly controversial.[15] We, nevertheless, should be clear about the sense in which they were representative. As already suggested, they were not necessarily representative of the American people or of a particular political party. They, rather, were representative of an influential type of American political thought at a critical historical moment when, in the crucible of debate, its assumptions were most fully exposed. They were the ones who most coherently articulated those assumptions. Their representativeness, then, was explanatory, not statistical. The comparison among this sample of American statesmen should serve to illuminate not only the contours of American political thought at three historical moments but also how those contours changed over time.

The comparison of particular cases raises the fourth "framing" question. What does it mean to call Webster, for instance, a republican? And what does it mean to say that he was less republican (or more pluralistic) than the Federal Farmer?

When applied to particular historical cases, republicanism and pluralism must be understood as ideal types. No American statesman was (is) purely the one or the other. Most were (are) somewhere in between the two. Their republicanism or pluralism is a matter of best fit.

This is familiar enough territory. Yet, there seems to be an irresistible temptation to treat republicanism as a historical force. The temptation is so great because it allows us to present the republican thesis as a causal explanation. Particular American statesmen were consciously committed to republicanism. This commitment explains (at least partially) why they held the particular ideas they did, and why those ideas changed over time as they and their intellectual heirs struggled to honor their commitment to republicanism under increasingly adverse social conditions.

With prodding, this "save republicanism" approach would be excused as a loose way of speaking and defended as a historical generalization necessary to elucidate a complex series of events. The problem with this approach is that it entails a gross overgeneralization of both the nature and the dynamics of American political thought.[16]

Not surprisingly, the "official" approach of republican revisionism is a more methodologically sophisticated, linguistic one. The more sophisticated revisionists, preeminently Pocock, begin with the truism that republicanism was an "essentially contested concept" that was defined in different ways by different American statesmen.[17] The essential contestability of republicanism (along with many of its constitutive concepts) suggests it is better understood as a multivocal political language than as an univocal historical force. The next step is to linguistically map the ideas of different American statesmen, disclosing how they, in particular, used the language of republicanism. (Because the historical actors themselves were embedded in a specific linguistic context, they were never fully conscious of how they were using the language.) Unless we make some highly unrealistic assumptions about the hegemony of political languages and the identity of our reconstructions of those languages with their own historicities, we cannot explain the ideas of particular statesmen in any causal sense. The goal, however, is not to offer a causal explanation but rather to offer an interpretive explanation which decodes their ideas in terms of their linguistic contexts.[18] Even so, we must make the same simplifying assumptions to go much beyond such crude insights as that: (1) certain political languages were available to American statesmen during certain epochs; (2) particular statesmen used those languages in particular ways; and (3) their usages led to a different set of linguistic possibilities in the future. The underlying problem with this approach is that the level of analysis is too general for it to provide a meaningful history of American political thought. It takes the focus off historical actors in such a way that we can only guess at what they actually thought, at what they intended rhetorically and what they proposed as a serious position. We then possess a history with only the vaguest notion of the nature of its particular cases and why they differed from one another, giving new meaning to the adage that the whole is greater than the sum of its parts.[19]

The obvious move at this point would be to pursue a mixed approach that somehow combines general and particular levels of analysis. But surely we should not adopt, as many republican revisionists implicitly have, an approach that merely goes back and forth between the two levels of analysis.[20] We should instead try to find an approach that is sensitive to both levels, a middle-range approach that isolates explanatory factors which are neither as general as political languages nor as idiosyncratic as personal motives. These factors, moreover, should be ones which strongly influenced most American statesmen across a significant stretch of time so that they can be used to help explain the differences between the political ideas of particular statesmen both within and between historical epochs. Assuming that nationalism is one such factor, our goal would not be to explain why Webster, to again use him as our example, was a nationalist. It would be to explain why his nationalism took such a different form than Calhoun's nationalism or, across time, than Publius's nationalism.[21]

The rest of this introductory chapter will be devoted to sketching a narrative of the history of American political thought constructed around three explanatory factors—traditionalism, social realism, as well as nationalism. I will suggest that the intersection of the positions particular American statesmen took on those factors locates them within ideological space and offers a powerful explanation of the differences between their overall positions. In subsequent chapters, I will then present, in much greater detail, three pivotal segments of this narrative.

American statesmen were, in the first place, committed to a particular tradition of political ideas, not to republicanism *per se.* In other words, they were committed to republicanism only insofar as we identify the nature of the prevailing tradition of ideas during their particular historical epochs as republican. Furthermore, they themselves were conscious of working within an evolving tradition of ideas; even when they used the word "republicanism," they did not think it implied an "either/or" commitment to an immutable set of ideas.

Only somewhat arbitrarily, I assume that in revolutionary America the prevailing tradition of ideas was purely republican.[22] In 1787–88, the political arguments of both the Anti-Federalists and Federalists were strongly influenced by their commitments to that tradition. It, however, was not exactly the same tradition; the social dislocations of the revolution had already spurred movement in a pluralist direction.[23] This evolution continued so that by the 1830s the Jacksonians and Whigs were committed to a still less republican tradition, and by the 1850s the Democrats and Republicans were no longer committed to an identifiably republican tradition but to a pluralist one. To speak precisely, the prevailing tradition of ideas had evolved from a predominantly republican to a predominantly pluralist tradition. Even today, a residual republicanism resists what has become an almost hegemonic pluralist culture.

The prevailing tradition of ideas steadily evolved in a pluralist direction because American statesmen were not merely traditionalists. They were also social realists who were committed to maintaining contact between their political ideas and changing social conditions. As the nation became more socially diverse or pluralistic, its prevailing tradition of ideas reflected that fact.[24]

Yet, this commitment to social realism does not lend itself to simple reductionist accounts. By the same token that American statesmen were not merely traditionalists, they were not merely social realists. They had moral intentions and those intentions were expressed through a commitment to a tradition of ideas which, at least initially, was profoundly normative and which never totally lost that original focus. They, thus, always appeared to be lagging behind social change, and often in disparate ways.[25] After all, neither the nature of social change nor of traditional ideas was unambiguous. There were seemingly endless permutations in how American statesmen could balance their dual commitments to tradition and social realism.

A third commitment pulled American statesmen, as a group, in the same pluralist direction as social realism. This was their commitment to American nationality and the union; in short, their nationalism.[26]

In 1776, federalism, both as a decentralized system of government and as an assemblage of smaller societies (states), was the undisputed answer to America's future as an independent nation.[27] This traditional answer, though, increasingly clashed with American nationalism. To minister to that tension by adjusting the prevailing tradition of ideas to a larger, national scale seemed to entail massive departures from that tradition. Predictably, such adjustments were resisted in the name of tradition, although never wholeheartedly because of the shared commitment to nationalism. The American statesmen who defended the more traditional, federal-republican position betrayed greater intellectual ambivalence than their less traditional, national-republican opponents did. They had the weaker argument. While they were not disunionists, it was relatively easy for their opponents to portray them as such. Moreover, their jeremiads in defense of the original federal republic lost much of their force when they had to admit that they did not exactly mean the republic of 1776, perhaps not even of 1787. Tradition and social realism, however, did appear to be on the side of the federal republicans. In rebuttal, the national republicans were compelled to argue that, under American conditions, their common intellectual heritage actually supported nationalism, not federalism, and that social change was centralizing, not decentralizing, the nation. These were not impossible arguments to make, but they were not as immediately persuasive as their contraries.

Ambivalence, thus, does not mean political impotence, nor does it mean intellectual incoherence. The federal-republican "package" on traditionalism, social realism, and nationalism was as tight (or loose) as the national-republican package. These packages brought some order to the permutations of possible positions on those three factors. The ratification debate was defined by the opposition between the federal-republican and national-republican packages and federal-republican and national-republican subtraditions soon emerged within a broader republican consensus. Yet, the continued interplay between traditionalism, social realism, and nationalism in particular American statesmen propelled the progressive development of pluralist elements within both subtraditions and, by the 1850s, a consensus shift. The Lincoln-Douglas debates evidenced the same opposition between states' rights and nationalism that had so dominated the ratification and nullification debates. That opposition, however, was now undergirded by pluralism, more than republicanism, and it was no longer definitive of political controversy in America. Lincoln and Douglas fundamentally disagreed over the meaning of pluralist democracy, not states' (or territorial) rights.[28]

Nevertheless, the gradualness, unintentionality, and incompleteness of this consensus shift bear emphasis. Even the most innovative, ultrarealistic nationalists, such as Publius and Webster, labored to update, not revolutionize, a particular, republican tradition of ideas. It is only from a historical perspective that we can observe their efforts as increasingly, and *seemingly* with a purpose, pushing that tradition in a pluralist direction. Furthermore, the American statesmen who at any point in time resisted "radical" revisions to that tradition participated in its transformation. They, too, were social realists and nationalists and they, too, acknowledged the need for some revisions. The Federal Farmer and Calhoun may have been less politically effective than their opponents, but the dynamics of intellectual change almost assured them of a measure of success. At least in politics, "conceptual revolutions" appear to be long, drawn-out affairs.[29]

On the one hand, the Civil War consummated the consensus shift from republicanism to pluralism within American political thought. The war was a catalyst for both social pluralism and American nationalism. It also, negatively, tainted the federal-republican (states' rights) subtradition, releasing the dialogical pressures on such unrepentant nationalists as Francis Lieber to justify their triumphant ideology on traditional, republican grounds.[30] On the other hand, republican revivals have periodically agitated the now-dominant pluralist culture. The Progressive era witnessed one such revival, with Theodore Roosevelt acting as the ghost of republican citizenship past.[31] The current proliferation of republican scholarship must also be fit into the narrative.

Republican revisionism is not "merely" history. The republican thesis is clearly didactic. It depicts the history of American political thought as a descent. A similar decline has allegedly occurred on the level of political practice, though this part of the tale is more cautiously told.[32] In a kind of cultural metempsychosis, the future possibilities of an improved, republican politics in America are implicitly attached to its prior incarnations.

To the extent that republican revisionism is republican revivalism, it reflects a widespread dissatisfaction with the pluralistic nature of contemporary American politics. This dissatisfaction has even affected Robert Dahl, presently the preeminent pluralist theorist. Republican revivalism also suggests that the nature of American politics is not completely pluralistic and that the evolutionary process I have just sketched is not a Weberian iron cage. The republican-pluralist dialogue within American politics is a continuing one because it is improbable that liberal statesmen would ever divorce interests from virtue, or virtue from interests. But does any one seriously doubt which will be the dominant partner in that relationship for the foreseeable future?[33]

As history, republican revisionism is also problematic. The republican thesis has been energized by didactic impulses which press for simplistic readings of the past to punctuate the lesson. It all too easily falls prey to facile tendencies to characterize American politics as so glorious "then" and so ignominious "now" and to dramatize the change, on a more theoretical level, as Mandeville's *Bees* infests Cicero's *Officis*.[34] What we need is a clearer recognition of how and why American politics has become more pluralistic, a recognition which this study begins to unfold.

2

The Categories of American Political Thought

In this definitional chapter, I will discuss the core beliefs of republicanism, liberalism, and pluralism. Republicanism and pluralism represent the poles of an evolving liberal tradition of ideas that centrally defines the history of American political thought. A discussion of these three categories of political thought is the necessary first step in any attempt to locate the political thought of particular American statesmen within that tradition.

I begin with republicanism because it was the predominant mode of American political thought during the late-eighteenth and early-nineteenth centuries. Once republicanism has been defined, it will be easier to show, first, how it is generically a form of liberalism and, second, how it differs from pluralism, the form of liberalism which became predominant in mid-nineteenth-century America. This discussion may appear unduly abstract and analytical. Yet, it is advisable in light of the imprecise ways republicanism has been used in the secondary literature, an imprecision which permits us to prove that every one, or no one, is a republican statesman.[1]

Republicanism

I will define republicanism in terms of sixteen core beliefs. These core beliefs constitute an ideal type which is "purer" than the political thought of any one American statesman. For purposes of historical explanation, however, I am assuming that the prevailing ideas of the American Revolution perfectly fit this ideal type. It will then serve as a guidepost for measuring how far subsequent bodies of American political thought evolved beyond the spirit of 1776. Of any one "work," the remarkable, fifty-year correspondence between Thomas Jefferson and John Adams probably best expresses that traditional republican spirit.

Even though the sixteen core beliefs of republicanism are closely interrelated, they can usefully be divided into four groups of four. These groups concern the ends of political society, the nature of citizenship, the aggregate qualities of the citizen body, and the character of the government. From these groups, it is clear that republicanism involves much more than the recommendation of a certain type of government. Republicanism is the wide-ranging public philosophy of a republican statesman. I will describe its core beliefs as a series of propositions to which he would assent.[2]

The Ends

§ The public happiness is the true end of any political society.

§ Liberty is also a fundamental political end. It is both an irreducible good and a necessary condition of happiness.

§ Public liberty takes precedence over private liberty. The right of citizens to participate in their collective self-government is not just another means of pursuing their own interests. Although it is importantly that, it is also an end in itself.

§ The public happiness can be defined organically, as a single good which subsumes the private interests of all the citizens of a political society and uniquely determines the optimal public policy in any given situation. Yet, the public good is also a construct of the long-term interests of the society. It, therefore, exists in some tension with the private interests and liberties of individual citizens. Those interests and liberties may have to be restricted in order for the society to pursue its collective interests.[3]

The inclusive nature of the public good, however, tempers this tension. Individual citizens will generally be free to pursue their own interests in their own way because that is in the public good. While far from absolute, private rights to life, liberty, and property are presumptively valid.

The basic republican strategy emerges here. It is to place the claims of public life over those of private life, without denying the validity of the latter. The second part of the strategy is to stress how much the two sets of claims overlap. Still, republican statesmen concede that a well-ordered society must sometimes demand stiff sacrifices of property, liberty, and even life from its citizens. These abridgements of fundamental private rights are not unjust, provided they are genuinely in the public good.

A republic, minimally defined as a political society which pursues its true, long-term interests, requires citizens willing to make sacrifices for that pursuit to continue. The next four core beliefs elaborate the nature of citizenship and what threatens and supports the development of good citizens.

The Citizens

§ Civic virtue is the essential quality of citizenship. In the first instance, civic virtue describes the willingness of individual citizens to sacrifice private interests to collective interests—the moral disposition which allows a society to pursue its true long-term interests without a despotic government. Second, civic virtue refers to the willingness of individual citizens to participate in public affairs, ranging from serving in government to discussing current events with their friends and neighbors. Good citizenship, though, is not excessively burdensome both because of the inclusive nature of the public good and, as we shall see, the reformative character of republican government.[4]

§ Moral declension is, nevertheless, inevitable. There is a constant tendency for the society's deliberate pursuit of its collective interests to be shunted aside by the citizens' frenzied pursuit of their private interests. Citizens are always inclined to fall back into the more natural state of unalloyed selfishness, to become wanton and licentious. And the failure of civic virtue is not merely an ever-present danger to republican societies; that is what historically has happened. Extraordinary virtue may call forth republics, but they never seem to survive ordinary selfishness.

Although republics are natural societies in the sense that they are consistent with the observed potentialities of human nature, they are also fragile societies because they exist near the outer edges of those potentialities. They push egoistic men to the limits of possible concern for a larger community. Other types of political societies are more stable since they do not depend as much on the moral character of their citizens.

History, thus, is cyclical. Republics inevitably degenerate into monarchies, or worse. Yet, there is some reason to hope that the cycle can at least be prolonged through the proper public policies.

§ Luxury overtaxes civic virtue. It accelerates moral declension by making citizens too attentive to their private affairs. The paths of civic virtue lead through more moderate (but, in line with the modern republican strategy, not Spartan) lives.

§ Republican government nurtures civic virtue by allowing individual citizens a considerable amount of public liberty. Because political participation and civic virtue are symbiotic, republican government contributes to the moral development of its citizens. (An alleged incapacity for the practice of civic virtue would, then, justify exclusions from the citizen body and perhaps even slavery.[5])

In republics, citizens must not only possess certain personal characteristics, they must also share certain aggregate characteristics. The third group of core beliefs defines the type of citizen body required in a republican society.

The Citizen Body

§ A highly homogeneous citizen body is a necessary condition of (relatively) stable republics. In the absence of social homogeneity, there can be no public good. If irreconcilable conflicts of interest develop between sizable groups of citizens, there can only be the tyranny of some (one) over the others. Ideally, republican societies are analogous to healthy organisms, in which all the parts of the body prosper together.

Of course, no political society is perfectly homogeneous and, again just as in the organic world, differentiation among the parts is highly advantageous from a developmental standpoint.[6] The question of how much social homogeneity, and in what respects social homogeneity, can only be answered in relativistic terms. Still, an equality of conditions is critical. A republic cannot be inegalitarian to the extent that class conflicts develop between a property-rich and a property-poor, for that will eventuate in either minority or majority tyranny. Class divisions also relocate virtue from the whole to the part. Finally, the specific vices of the rich—arrogance and wantonness—and of the poor—envy and licentiousness—run riot.

To once again anticipate later discussion, republican governments need not be particularly active, nor should they be particularly passive, in preserving republican equality. Unrestrained pursuits of wealth will create indigestible economic inequalities because men are naturally unequal. Virtuous citizens, though, will not excessively seek wealth and the economic inequalities that result from properly moderated pursuits of wealth are not only legitimate but socially desirable. Republican governments, therefore, are left with the task of containing private enterprise within fairly broad channels of economic activity. They normally do not have to pursue leveling policies or enact sumptuary laws.

§ Republics must be small. The argument for small republics is one of comparative advantages. The smaller the republic, the more likely the citizen body will be highly homogeneous, the closer the government will be to the people, and the more strongly the civic virtue of individual citizens will be reinforced through political participation. In the smallest republics, every citizen can directly participate in government as a member of a democratic assembly. But beside the fact that direct democracies are not the most desirable political societies (as will be shown below), they also are totally impractical under modern conditions. Size has its obvious advantages. Larger states are more powerful ones, both in their military and economic capacities.

A corollary to the small-republic argument, then, is to recommend the formation of federal republics as a way of combining the comparative advantages of smallness and size.[7] Exactly how power should be distributed between the governments on the two levels of a federal system is open-ended as long as the (member-) state governments remain the more powerful ones. Generally, this imperative requires the federal government to act through the state governments, not directly on individual citizens. In this manner, the locus of public life for most citizens will remain the small-scale, state republics and the federation will be a "republic" by aggregation.

§ Commerce destabilizes republican societies. It introduces a complex economy, thus undermining social homogeneity. Even if it does not produce class divisions, it will inevitably increase economic inequalities. By the same token, it decreases the likelihood of a collective good which subsumes all significant economic interests. It, finally, saps civic virtue by spawning luxury and, on the opposite end of the economic spectrum, propertylessness. Commercial republics are invariably short-lived.

However, commerce need not be shunned altogether in republics. It is neither possible nor expedient for modern societies to be agricultural societies. Commerce only has to be pursued with moderation. This injunction holds both on the aggregate level—where a majority of the citizens should be independent yeomen—and on the individual level—where the citizens engaged in commerce should act as if they were independent yeomen. There is a definite pecking order on the civic safety of various commercial occupations. While a merchant or an artisan maintains something of the personal property, economic independence, leisure, and, hence, virtue of the yeoman farmer, a mechanic employed in a "manufactory" does not. Similarly, foreign commerce is beneficial for exporting agricultural surpluses but not for importing luxury items. If such distinctions consistently inform public policy, commerce can actually be a blessing to republican societies in significantly boosting public wealth.[8]

§ Political parties are conspiratorial. Their *raison d'être* is ensuring that government policies pursue the narrow interests of their members, usually at the expense of the interests of the larger community. They also deflect the virtue of their members from the whole to a self-serving part of the citizen body. (Needless to say, the evaluation of political parties formed to pursue certain broad principles of public policy or, as will be discussed shortly, to subvert despotic governments is much more positive.)

Some of the same tensions clearly exist between interest groups and the larger community, and they can be lumped with political parties under the generic term "factions." Yet, interest groups, unlike political parties, are endemic to modern societies and their very multiplicity makes them less dangerous.

Politics, paradoxically, is both an elevating and a corrupting form of human activity. A republican government is designed to stimulate the former, and depress the latter, tendencies. The final group of core beliefs discloses the nature of this wondrous type of government.

The Government

§ Republican governments are popular governments.

§ Republican governments are also mixed governments.

These beliefs will be discussed, at some length, together because it is the mixture of democratic and aristocratic elements in republican governments which distinguishes them from democratic governments. These two types of popular government are invariably confused.[9]

Strictly speaking, republican and democratic governments are distinguished by the representative principle. Each citizen can directly participate in a democratic government. But, as already suggested, direct democracy is neither ideal nor practical under modern conditions. Representation is not just a necessary evil, forced on republics by size. It is a positive good.

Republican and democratic governments, however, can also be distinguished as two types of representative governments. Four criteria will be used to bring out this distinction: (1) who elects various public officials; (2) how many are elected to each office; (3) how often are they elected; and (4) what kind of person is elected.

According to these criteria, democratic governments possess the following characteristics: (1) all adults can vote for all major public officials; (2) the number of people elected to each office, especially to the legislature, is at the upper, practical limits; (3) they are frequently elected; and (4) a representative sample of the adult population is elected.[10]

Republican governments diverge from democratic governments on these criteria because it is assumed that less democratic means are more likely to assemble a government that systematically pursues the public good. Most important, a representative sample of the adult population should *not* be elected to government. There, instead, should be an overrepresentation of "the best and the brightest" in government. To prejudice that result, variations are, in turn, advisable on the first three criteria. At least for some major offices, the electors should be severely restricted, a relatively few people should be elected, and they should be elected for relatively long terms.

These criteria establish a continuum of representative governments that are more or less democratic. The American revolutionaries staked out a decidedly democratic-republican position. They rejected hereditary monarchy and aristocracy and they favored republican governments which included: (1) broad-based electorates and (minimally) the indirect election of all major officeholders; (2) a large, directly elected legislative body; (3) a frequently elected legislative body; and (4) a mix of "democratic" and "aristocratic" representatives.[11]

For republican statesmen, democracy and aristocracy are not merely two abstract principles of government. They are also social principles which define two kinds of people. They refer to a natural hierarchy between people of average and superior talents; between natural democrats and aristocrats. Both kinds of people are needed in republican governments, but for different reasons. The role of the democratic representatives is to secure the proper sympathy between the government and the governed. Their role is largely negative, to prevent government policies from pursuing the interests of only a narrow elite. Aristocratic representatives perform a more positive role. They promote the public good as the conscious aim of government policies. They possess a firmer vision of the true, long-term interests of the society and of how to best pursue those interests through government policies. They also are more willing (and able) to endure the personal sacrifices often demanded of public officials in the course of that pursuit. Where democratic representatives mirror the citizen body, aristocratic representatives refine it.

Although the natural democracy and aristocracy are essentially defined by a hierarchical difference in abilities, they are phenomenally two social orders roughly measured by their relative affluence. On these grounds, the analogy between republican governments and the traditional, English model of mixed government is particularly strong—all the more so if separate legislative bodies are retained for each order. Yet, the two models of mixed government differ precisely to the extent that natural orders differ from hereditary ones. The former are less steep, more fluid, and, therefore, do not violate a republican equality of conditions. A republican government, unlike the English government, is completely open to talents, regardless of family background.

The new, republican model of mixed government is an appropriate subject for institutional experimentation.[12] This experimentation must take into account the countervailing tendencies among citizens to vote for candidates just like themselves and to vote for candidates who they acknowledge as their superiors in some respect. Ultimately, though, a republican government is more analogous to a democratic government than it is to a traditional, mixed government insofar as its internal mixture can, and should, be attained through mostly democratic means.[13]

In sum, republicanism is a democratic teaching. The dominant message is equality. The recognition of certain inequalities as both inevitable and desirable should not obscure that message. Republican statesmen qualitatively distinguish the benign economic inequalities which are based on a natural diversity of talents from the cancerous ones which are not. They insist that the organization of a government should not ignore natural inequalities; but, just as important, it should not aggravate artificial ones. They show considerable faith in the capacities of the average citizen. Given the proper social and political arrangements, he or she is expected to practice civic virtue, to actively participate in politics, and to wisely vote for a variety of public officials. Republican statesmen also assume that the average citizen can rise to the demands of good government, in part because of the way government service elicits the very qualities it demands of him or her. Even though these beliefs do not rebut the argument for the overrepresentation of superior talents in government, the implications are still strongly democratic.

Republicans value a widespread distribution of citizenship rights, both for its own sake and for its beneficial effects on government. The sense in which they think government service does *not* nurture civic virtue strengthens this commitment to public liberty. The implications of the last two core beliefs of republicanism are also strongly democratic.

§ All power corrupts. Because all possible gradations of power, public and private, are breeding grounds of corruption, moral decay incessantly eats away at the foundations of republican societies. This tendency compels periodically "returning to first principles."[14]

Nonetheless, the dangers different developments pose to republics are still distinct in nature and epidemiology. Government corruption is the more immediate threat than luxury or commerce. The temptations of office soon ensnare even the most superior virtues. Once more, the moral effects of political activity appear paradoxical. While government service improves character for a certain period of time, it then "all too suddenly" becomes unsalutary.

The development of a court party is the most ominous threat to republican societies. A court party is the worst type of faction because it is one entrenched in power. Through its official status, it can systematically skew the pursuit of the public good toward the collective interests of government officials and their dependents. A court party matures into an artificial aristocracy whose wealth and power far surpass any natural justifications. Despite the nomenclature, a hereditary monarch (or peerage) is not essential to its existence. It can be organized and sustained entirely through elective means. A shrewd manipulation of government patronage and revenues can subvert free elections without destroying their forms.

The antibody of a court party is a country party. A country party is a counterorganization of the "independent" part of the citizen body. It is a party temporarily called into being by a successful conspiracy against the republic. As such, it is actually a testament to the evils of party. If the "Country" fails to purge the government of its corrupt court influences, then a revolution is the remedy of last resort.[15]

The institutions of a republican government, however, act to forestall these worst-case scenarios. Frequent elections, for example, afford vigilant citizens the opportunity to set up an effective rotation in office, quietly retiring government officials who are on the verge of corruption.

§ Republican jealousy is an important component of good citizenship.[16] Due to the strong correlation between political power and corruption, ordinary citizens should be suspicious of government officials. The distinctive action which results from republican jealousy is voting out of office any official with a tincture of corruption.

Beside frequent elections, other institutional supports of republican jealousy are advisable. A written constitution that formally limits the powers of government officials is a key auxiliary precaution. It provides a standard of judgment for suspicious citizens to apply to the conduct of their government representatives. A republican constitution may also institutionalize the rule of law by establishing a separation of powers between those who make, execute, and interpret laws. Again, the precise blueprint will vary from society to society.[17]

Any recommendation of limited government, though, is a weak one, as is true on the policy level of *laissez faire*. Government officials must have powers and there are obvious advantages, as well as dangers, to their actively using their powers in pursuit of the public good. Just as a certain amount of republican jealousy of public officials is necessary to suppress the dangers, a certain amount of popular deference toward public officials is necessary to reap the advantages.

Before concluding this discussion of the core beliefs of republicanism, I should mention what might be termed its metabelief. The citizens of a republic must themselves share these sixteen core beliefs. Republican statesmen only do so more reflectively and articulate them as a public philosophy.

Liberalism

By utilizing these sixteen core beliefs to define an ideal type of republicanism, we can greatly clarify the historiographical debate over the nature of American political thought during the pre–Civil War period. We can also clarify that debate by separating the core beliefs which are genus-liberal ones, and thus common to all forms of liberalism, from those which are specific to republicanism. As I noted in chapter 1, there is a tendency on both sides of the historiographical debate to overlook the genus-liberal beliefs in republicanism and, consequently, to claim that it is a nonliberal, even a neoclassical, category of political thought. Equally, there is a tendency to conceive of liberalism in narrowly atomistic terms which define only its most extreme pluralistic forms.[18]

The early-modern philosophers understood themselves as initiating a radical break with classical and medieval philosophy. The opening paragraph of chapter fifteen of Machiavelli's *Prince* is the *locus classicus* of this break.[19] Following Machiavelli, philosophers such as Hobbes and Locke reassessed the relation between politics and other types of human activity. When they had completed that reassessment, politics was less autonomous from activities which had previously been considered "lower"—principally economics—and more autonomous from such "higher" activities as religion and philosophy. Aristotle's treatment of household management in *The Politics* and Locke's treatment of property in *The Second Treatise* are worlds apart.[20] Depending on one's perspective, this emergent liberalism meant the unfortunate (and unnecessary?) reduction of politics to the cave or else its providential expansion to encompass new social forces. From either perspective, it reflected a greater emphasis on drawing boundaries between public and private activities, an emphasis which could equally protect the public sphere from immoderate private passions as protect the private sphere from stultifying community norms. This emphasis was felt to be more consistent with the conditions of modern societies and, perhaps more important, with the potentialities of human nature. The original liberal philosophers criticized older modes of political thought both for being antiquated and for never having been realistic in the first place.[21]

Liberalism, therefore, centrally defined itself in relation to classical and medieval philosophy. The original liberal philosophers were united by a distinctively modern vision of the human ends, virtue, society, and government.[22]

The Ends

For liberals, liberty and happiness are the fundamental ends of any political society.[23] Classical and medieval philosophers embraced other ends, such as the cultivation of human excellence or religious faith. Liberals see those ends as private, not public, ones.

The liberal ends identify a certain way of life as the most desirable one, aptly phrased as "commodious living."[24] In liberalism, happiness has primary reference to material interests, liberty to rights to pursue such interests, and justice to the protection of such rights. Even though the liberal presumption in favor of private rights to life, liberty, and property is rebuttable, that still is a presumption which was not shared by earlier philosophers. Nor, did they accord the same sanctity to how individuals defined their own happiness or interests.

The Citizens

Liberals sharply distinguish political virtue from the moral and intellectual virtues.[25] They, furthermore, view all the virtues as means to happiness, not as ends in themselves. In contrast, earlier philosophers valued happiness as a consequence of the fully virtuous life. Their definitions of citizenship were the more demanding ones. But, then, they thought that only a few could reasonably aspire to be good citizens and that only those few should be granted citizenship rights.

Of course, liberals claim that, as a matter of fact, political virtue overlaps the other virtues and that someone can not truly be happy without cultivating all the virtues. The contrast with their predecessors, though, clearly emerges in the different rhetorics they use when virtue and happiness do not seem to coincide. Rather than arguing that some action which does not appear to be in the actor's interests is, in any case, the virtuous thing to do, liberals attempt to persuade the actor that it really is in his or her interests.

The Citizen Body

Liberals believe that a perfectly homogeneous society, even if possible, would not be desirable. As compared to their predecessors, they subscribe to a looser definition of what holds a society together. Perhaps all that is required is an agreement on liberal principles. Liberals also see more benefits in diversity, as in how it insulates individual citizens from societal pressures. Their more relaxed sociology includes more favorable attitudes toward commerce and size.[26]

The Government

While not necessarily popular governments, liberal governments are nevertheless relatively democratic ones.[27] The personal abilities required to perform various political roles are, across the board, less rigorously defined in liberalism than they were in classical and medieval thought. The dissociation of political from philosophic (or prophetic) wisdom and the expansion of citizenship rights intensify the democratic thrust of liberalism. Although liberals tend to hold a less hierarchical view of human nature than their predecessors did, the more significant difference is their less hierarchical view of the demands of public life. Liberals, however, do not necessarily advocate greater economic equality. Indeed, a disjunction between political and economic (in)equality is highly probable in liberal societies because of the manner in which they segregate human activities into public and private sectors.

Pluralism

In a way akin to the early-modern liberal philosophers, the early twentieth-century pluralists understood themselves as a vanguard for a more realistic political philosophy or, now even more, science. They insisted that their predecessors were irrelevant to the new social order which had risen from the ashes of the Civil War and seriously questioned how relevant they had ever been to the course of American political development, given their narrow political focus, value-laden language, and nondemocratic intentions. Arthur Bentley and Charles Beard naturally did not identify their targets as liberal republicans, nor their critiques as critiques within liberalism. Yet, it was precisely because earlier American political "scientists" were liberals that they were so vulnerable to the pluralist assault.[28]

Predictably, the early pluralists emphasized how much they differed from their republican predecessors. This emphasis obscured their common liberalism. Pluralist and republican statesmen agree that liberty and happiness are the fundamental human ends, that some kind of political virtue and some degree of social homogeneity are prerequisites of liberal societies, and that only popular governments can be good governments. They, nonetheless, do disagree in many respects.[29]

The Ends

Pluralists reverse the priority republicans give to public over private liberty. According to pluralists, citizenship rights provide an important venue for the pursuit of private interests "by other means"; they are not ends in themselves.

In pluralism, there also is no public good, at least not as republicans conceive of one. Whatever public good there is, is a summation of individual or group interests. Pluralists analyze society into the relations between and among individuals and groups, not individuals and *the* society.[30] The public-policy process, therefore, becomes a matter of bargaining over group interests instead of deducing the optimal public policies from some abstract notion of the true, long-term interests of the society.

Relative to their republican predecessors, the early pluralists carried out both a major redefinition of the boundaries between the public and private sectors and a substantial shift in emphasis between the two. Pluralists share a stronger presumption in favor of private rights, and they more tightly associate justice with the protection of those rights. Just as in the case of the original liberal redefinition of the public-private boundaries, the end result of the pluralist redefinition of those boundaries was equivocal. In one sense, the pluralists expanded the boundaries of the public sector by arguing that human activities which had previously been considered private were actually public. This "insight," after all, formed the basis of their interest-group paradigm of politics. In another sense, however, they expanded the boundaries of the public sector only to devalue the activities which had previously been considered quintessentially public. Public debate, for instance, was now dismissed as the mere shadow play of real political conflict.

The Citizens

Pluralists identify political virtue with fair play. (In pluralism, game metaphors replace the organic metaphors of republicanism.) Individual citizens must be willing to abide by the rules of the game. Citizen "sacrifices" are accepted losses, not supererogatory acts. Pluralists are also less likely than republicans to interpret political and economic conflicts as zero-sum games. They, thus, are more accepting of conflict and less censorious of citizens' spirited assertion of their own interests. All in all, the pluralist conception of political virtue is not nearly as stringent as the republican conception.

These different conceptions of political virtue are directly linked to several other differences. Unlike republicans, pluralists do not especially fear moral declension. They hold a more progressive view of history; or, just as likely, their public philosophies are essentially ahistorical. They also do not dread luxury. To the contrary, they feel that luxury can be a valuable incentive to spurring economic growth. Pluralists, finally, do not assume that a symbiotic relationship exists between political virtue and political participation. Pluralist societies and governments do not, and should not, seek to produce better citizens.

The Citizen Body

Pluralist societies, of course, are not highly homogeneous. Accordingly, pluralist statesmen are less ambivalent toward social diversity than their republican predecessors were; conflict, not consensus, is the primary reality in their social vision.

Still, pluralists acknowledge the need for some degree of social homogeneity or consensus. Minimally, there must be widespread agreement on the rules of the game in pluralist societies. Beyond that, the absence of bipolar racial, ethnic, or religious divisions is almost essential. Economic inequalities should also fall within certain limits, though much broader ones than republicans would think admissible.[31]

In contrast to republican societies, pluralist societies are under no size constraints. They are indefinitely expandable and can enjoy all the comparative advantages of size without having to federalize. Within the context of a federation, however, the differences between pluralists and republicans would be most visible on the state level. Pluralist governments are, in general, similar to the weak central governments of federal republics and dissimilar to the more active state governments in such regimes.[32]

Not surprisingly, pluralists are also less ambivalent than republicans toward commerce. In fact, pluralists cheerfully face the commercial future, heralding its wealth-producing potential and discounting its potential for social divisiveness. They do not morally typecast different occupations, nor do they qualitatively distinguish between different groups of citizens engaged in the game of politics as long as they play by the rules. Joining a political party or an interest group is one strategy—and an obviously rational one—which citizens should feel free to pursue in attempting to maximize their own interests. Pluralists, unlike republicans, are not tempted to view politics in conspiratorial terms.

The Government

Both pluralist and republican governments are popular governments. They, nevertheless, are based on different theories of representation. The pluralist theory of representation further stretches any analogy to traditional models of mixed government.

Pluralists believe that popular representatives should assert the interests of their political parties or interest groups in the public-policy process. While pluralists realize that being a good representative requires some special game-playing skills and moral habits (to remain a faithful representative), those qualities embody a lesser wisdom and virtue than republicans consider necessary in a good representative. Pluralists, then, place an increased emphasis on the sympathetic and social-mirroring functions of representation and a reduced emphasis on the refinement function.

These different theories of representation do not suggest opposing views of human nature so much as they suggest opposing views of the public-policy process. Contemplating only process-defined policy goals, pluralists do not see the same need for a natural aristocracy in government. They analogize pluralist democracy to a free-market economy, arguing that the less statesmen try to consciously guide the democratic process, the better it works. As a result, pluralists can fully indulge in the elitist dangers inherent in such efforts and vigorously defend majority rule, which they claim is, in practice, minorities rule anyway.

Despite their greater acceptance of economic inequality and their weaker commitment to public liberty, pluralists are the stronger democrats. They would be closer than republicans to the (pure) democratic position on the four criteria of popular government discussed above.[33] They recognize less need for political hierarchy. Their social vision is also more democratic. Even as they "discover" a more complex social reality, they insist that it is a less stratified one. In short, pluralists reject most of the presuppositions of mixed government.

Sharing similar views of human nature, both republicans and pluralists believe that all power corrupts. Pluralists, though, consider that tendency a more innocuous one. They do not perceive the same abyss between a normal and a corrupt public-policy process. Alternatively, even a well-functioning pluralist government would be corrupt from a republican perspective. The pluralists' more realistic definition of political normalcy entails the elimination of such "emotive" terms as the public good, factions, and tyranny from their vocabulary.

According to pluralist statesmen, ordinary citizens should be suspicious of government officials. But, once again, this disposition has different connotations within pluralism and republicanism. Pluralists counsel citizens to act to ensure that *their* representatives remain faithful to *their* interests, not to an overarching public good. For their part, republicans would disavow the pluralists' whole interest-group paradigm of politics.

Pluralism does stand as the more realistic pole of American liberalism. It is locked into major trends that have characterized America—a nation conceived in modernity—from its inception. Republicanism, conversely, stands in opposition to modern trends toward the privatization of human existence, commerce, luxury, social diversity, nationalism, political-party systems, and, to a lesser degree, democracy. This conclusion, however, is not unambiguous.

For one thing, republican and pluralist statesmen hold different views of the relation of thought to reality. Pluralists offer their conception of politics as an empirical one, perhaps strictly so, while republicans are more self-consciously normative.[34] Thus, the latter, if confronted with the choice, might concede that pluralism is the more realistic public philosophy during a particular historical epoch without, *ipso facto*, abandoning their republican beliefs.

The second ambiguity lies within the very criterion of realism. Even putting to one side the inherent subjectivity of the criterion, we can reasonably assume that it will not uniquely define one conception of politics as realistic during a particular historical epoch. American statesmen might, then, agree that pluralism was more (or less) realistic than republicanism during a given epoch and yet also agree that both conceptions of politics fell within an indeterminate range of credibility. In these terms, the republican thesis is that in 1776 most American statesmen, notwithstanding a certain skepticism about their own and other Americans' motives, considered republicanism a credible as well as a desirable public philosophy but that they soon lost faith in its credibility if not also in its desirability.[35]

The story is more complicated still. Even the traditional republicans of 1776 were not really opposed to modern trends. What they were opposed to, from their eighteenth-century perspective, was the further radicalization of those trends; for instance, the transformation of home industry into manufactories.[36] Their opposition proved ineffectual not only because modernity had its own momentum but because they did not possess the intellectual resources with which to resist that momentum. They could only adapt to it, which they and their successors did in such a way that the American political tradition, as a whole, became increasingly pluralistic. These dynamics, though, were not just imposed on, and subversive to, that tradition. They were also internal to it, because it was a liberal tradition. The gradual abandonment of republicanism for public philosophies more congenial to modern trends was prefigured at the origins of liberalism when it was associated, in spirit if not always in detail, with those trends.

Where traditionalism retarded the shift from republicanism to pluralism in America, nationalism accelerated it. At any given time, America looked less like a traditional republic as a nation than it did as a federation of smaller, more homogeneous states. To be credible, nationalist statesmen had to be more pluralistic than their federalist or states-rights rivals. Traditionalism and nationalism, however, were not simply conflicting ideological forces, any more than traditionalism and social realism were.

After all, what was at stake to most nationalist and federalist statesmen in 1787 was the American republic, not a particular tradition of ideas. The nationalists of that era—confusingly called Federalists—argued that America could only remain united as a national republic, even if that state of affairs ill-fit traditional republican ideas. More boldly, they argued that America would make a better republic as a nation than it had, or would, as a federation. For them, the American republic was the union. These claims were difficult to rebut both because the Anti-Federalists (really federalists) were also ideologically committed to the union and because their appeals to the sanctity of a certain tradition of ideas proved illusory, especially when they, too, were unwilling to treat it as something sacred. On the one hand, they proclaimed the states the American republics. On the other hand, they, like their opponents, were enchanted by the vision of a powerful, missionary republic and would do whatever they thought was necessary to save the vehicle of that vision. Ultimately, the Anti-Federalists hedged their bets on the future by advocating a "partly national, partly federal" republic. Their position, no matter how reasonable it may have been, was rhetorically the more vulnerable one.[37]

A central ambiguity, therefore, ran through the ratification debate: Was the preservation of the American republic and the union the same political act? Only by the end of the Civil War was the answer clear—or, at least, no longer disputed.

PART ONE

THE RATIFICATION DEBATE

The debate over ratification of the United States Constitution crystallized the partially conflicting commitments of the American founders to traditionalism, social realism, and nationalism into two general theoretical positions: national and federal republicanism. These positions underlay support for and opposition to the Constitution. The whole debate came to pivot on the question of whether the union or the individual states would make better republics. The irony was that the founders who answered "the individual states" did not seem to think any less of the union. In part 1, I will analyze the political thought of Publius and the Federal Farmer as representatives of, respectively, national and federal republicanism during the founding era. My controlling assumption is that we can better see the dynamics of, and between, those two positions in them than in other Federalists and Anti-Federalists. This analysis will crucially depend on an assessment of their places within an evolving American liberal tradition, a task for which we should be well prepared by the discussion of the preceding chapter.

3

Anti-Federalists and Anti-Republicans

In the first of the eighteen public letters he wrote opposing unconditional ratification of the Constitution, the Federal Farmer outlines three general plans of government for the new nation. After rejecting the first, federal plan, he makes a startling admission.

> The third plan, or partial consolidation, is, in my opinion, the only one that can secure the freedom and happiness of this people. I once had some general ideas that the second plan [of complete consolidation] was practicable, but from long attention, and the proceedings of the convention, I am fully satisfied, that this third plan is the only one we can with safety and propriety proceed upon.[1]

This statement is startling both in view of the tenor of the rest of the Federal Farmer's letters and of the way he and other Anti-Federalists have previously been interpreted. It suggests that they distinguished themselves from the Federalists by their political caution, not by their lack of nationalism. An uneasy combination of political caution and nationalism, however, turns out to be the key to the Federal Farmer's political thought and, I would argue, the Anti-Federalist position.[2]

That combination explains why the Federal Farmer should appear as a lukewarm opponent of the Constitution. Even though he considers it a plan of complete consolidation, he commends the "many good things" in the Constitution and contends that it provides "a better basis to build upon than the confederation"(FF, p. 252, letter V; p. 257, letter VI). He merely wants it amended in such a way that it assumes more of the features of a plan of partial consolidation (see FF, pp. 255–58, letters V–VI).[3]

If the Federal Farmer's political caution and nationalism together define partial consolidation as the best, practical plan of government, then what defines the impracticability of the ideal plan of complete consolidation? Clearly, it is the incongruence he perceives between that plan and American social conditions. His political caution takes the form of closely following a relativistic logic which fits plans of government to social conditions. According to this logic, the Constitution is indefensible.

The Federal Farmer accuses the proponents of the Constitution of overlooking the question of fit. In a probable reference to Publius—"the lengthy writer in New-York"—he criticizes "his pieces" for having "little relation to the great question, whether the constitution is fitted to the conditions and character of the people or not" (FF, p. 306, letter XIII).[4] Yet, it is not the lack of fit *per se* which he finds so objectionable. It is a lack of fit that is likely (calculated?) to destroy conditions which are highly desirable from a certain moral point of view.

> If there are advantages in the equal division of our lands, and the strong and manly habits of our people, we ought to establish governments calculated to give duration to them, and not governments which never can work naturally, till that equality of property, and those free and manly habits shall be destroyed; these evidently are not the natural basis of the proposed constitution. No man of reflection, and skilled in the science of government, can suppose these will move on harmoniously together for ages, or even for fifty years. (FF, pp. 251–52, letter V)

At times, the Federal Farmer portrays the Federalists as simply deficient in the science of government. At other times, he portrays them as skillful political scientists whose "errors" are malevolent ones (see FF, p. 253, letter V; p. 258, letter VI). He, not surprisingly, presents himself as someone who is sound on both technical and moral grounds. From his perspective, the best—in the dual sense of best—political science of the day is the small-republic argument. It dictates that America remain a federal republic. It becomes his central theoretical argument against ratifying a plan of government he thinks will consolidate and, thus, undermine the American republic. So tight does he see the association between federal and republican principles that he claims the Anti-Federalists and Federalists should rather be called republicans and anti-republicans (see FF, pp. 258–59, letter VI).

Nevertheless, the Federal Farmer advocates a plan of partial consolidation. Despite the traditional arguments and sociological postulates which oppose an American national republic, he believes that American conditions, to some extent, do support such a republic. Given the dynamic character of the fit question—after all, he clearly thinks the wrong type of government can produce undesirable social change—he is even willing to contemplate greater consolidation in the future. Indeed, he seems to wish American political development would take that course. What becomes crucial is that he fears the undesirable more than he dares the desirable. He is open to experiments in new types of republican governments and societies. The Constitution, though, is a rash experiment to him. It goes too far, too fast, toward consolidation, neglecting the fragility of republican societies and foreclosing the options of succeeding generations to denationalize American politics if it proves insufficiently republican on that scale (see FF, pp. 239–40, 243, letter III; p. 336, letter XVII).

The Federal Farmer's political thought was a compact package of normative beliefs, empirical observations, and methodological cautions. Primarily, it was a traditional federal-republican package. Stated in a way more faithful to its nuances, it was a "partly national, mostly federal" package with significant nontraditional elements. What we gain from the more subtle description—beside a certain awkwardness of expression—is the recognition that his thought represented a fairly successful synthesis of several conflicting theoretical commitments. This view seems superior to the prevailing interpretation of Anti-Federalist thought as simply (chiefly?) a reflection of a hegemonic republican ideology or language. At issue, of course, is not only the coherence of his political thought but the adequacy of our understanding of it; or how adequately we want to understand it.[5]

A Question of Fit

Like any good republican statesman, the Federal Farmer dwells on the social prerequisites of a republican government. The dominant strain in his social analysis is that the diversity of the American empire requires a strongly federal system of government for it to remain a republic. American conditions are most realistically perceived as defining a federation of smaller, state republics. This perspective mandates complete rejection of the Constitution as a consolidated plan of government.

However, the Federal Farmer's social analysis also betrays a nationalistic undercurrent that defines America as something more than a federal republic. This set of observations allows for the realistic possibility of a national system of republican government in America; for the present in muted forms but perhaps in stronger forms in the future. From this point of view, the Constitution is merely premature.

Traditional Views: America as a Federal Republic

The Federal Farmer claims that all possible sources of authority—the American people, great political authors, historical experience, and even his opponents—agree that extensive republics must be federal ones. That agreement defines a tradition. He feels it is not even necessary to argue such a truism.

> I believe the people of the United States are full in the opinion, that a free and mild government can be preserved in their extensive territories, only

> under the substantial forms of a federal republic. As several of the ablest advocates for the system proposed, have acknowledged this . . . I shall not take up time to establish this point. (FF, pp. 330–31, letter XVII)

But, of course, he already has.

To the Federal Farmer, "the first interesting question" suggested by the Constitution is "how the states can be consolidated into one entire government on free principles" (FF, p. 228, letter I). The resounding answer of his letters is that it cannot be done. He argues that it cannot be done in three different senses of republican government: as a government which protects private liberty; as a government which actively pursues the public good; and as a government which possesses a full and equal representation.

In the first place, the Federal Farmer argues that in America a consolidated system of government cannot be based on principles of private liberty. Political geography turns out to be a necessary, although certainly not a sufficient, condition of republican government even in this highly consensual, liberal sense of free government.

At the end of the first letter, the Federal Farmer raises the classic objection to an American extended republic.

> Independant [*sic*] of the opinions of many great authors, that a free elective government cannot be extended over large territories, a few reflections must evince, that one government and general legislation alone, never can extend equal benefits to all parts of the United States: Different laws, customs, and opinions exist in the different states, which by a uniform system of laws would be unreasonably invaded. (FF, p. 230, letter I)

The Federal Farmer's initial formulation of the small-republic argument is a crude one. He argues as if distance *per se* were the determining factor in the willingness of citizens to obey a government and, thus, in its capacity to be a free government.

> There are other considerations which tend to prove that the idea of one consolidated whole, on free principles, is ill-founded—the laws of a free government rest on the confidence of the people, and operate gently—and never can extend their influence very far—if they are executed on free principles, about the centre, where the benefits of the government induce the people to support it voluntarily; yet they must be executed on the principles of fear and force in the extremes—This has been the case with every extensive republic of which we have any accurate account. (FF, p. 231, letter II)[6]

Here, the Federal Farmer invokes the broad distinction between governments which essentially rely on persuasion for citizen compliance and those which essentially rely on force. As he later develops this distinction, it is the distinction between free and despotic governments (see FF, p. 264, letter VII). He contends that the critical factor separating these two types of governments is citizen confidence.

> The great object of a free people must be so to form their government and laws, and so to administer them, as to create a confidence in, and respect for the laws; and thereby induce the sensible and virtuous part of the community to declare in favor of the laws, and to support them without an expensive military force. (FF, p. 234, letter III)

According to these distinctions and definitions, the Constitution stands condemned by the following logic: (1) one government, operating over a large territory, must be despotic because it is too far removed from most citizens for them to place their confidence in it and voluntarily obey its laws; (2) under the Constitution the federal government will, for most important purposes, be one government operating over a large territory; (3) it, therefore, will be a despotic government (see FF, pp. 233–34, letters II–III; p. 265, letter VII).[7]

The obverse side of this logic, where size is the crucial societal variable, is that the state governments can be free governments because they each operate over much smaller territories. The Federal Farmer never seems to doubt that the state governments *are* free governments which have attracted popular confidence and obedience (see FF, p. 233, letter II; pp. 338–39, letter XVII).[8] As a result, a federal plan of government, which, contrary to the Constitution, leaves the state governments more powerful than the federal government, is necessary to free government in America.

In addition, a fairly specific division of powers in a federal system follows from the force-persuasion distinction. The Federal Farmer claims that those powers of government most intimately connected to the use of force should be exercised by the state governments. These are the internal police powers; strictly speaking, the administration of criminal and civil justice. Insofar as the state governments have collected the popular confidence, they will be able to minimize the actual use of force in administering justice. The case would be far different if the federal government administered justice, as it cannot count on the same degree of citizen confidence.

The Federal Farmer (wrongly, in the long run) does not expect the federal government to administer justice to any significant extent, though he is troubled by the potential contained in interstate diversity and bankruptcy cases (see FF, p. 243, letter III; pp. 344, 347, letter XVIII).[9] He, nonetheless, believes that other powers of government may have a similar impact on its fundamental character. On analogy, they can be called internal police powers. The analogy is particularly strong with respect to the powers of taxation.

Under the Constitution, the federal government will be able to directly tax individual citizens even though, according to the Federal Farmer, most of them will probably not be inclined to voluntarily comply with its tax laws. He predicts that the federal government will need an "army" of revenue officers to collect its taxes (see FF, pp. 239–40, letter III; pp. 332, 337, letter XVII).[10] When he urges returning the internal police powers to the states, his primary reference is to the powers of the purse (see FF, p. 229, letter I; p. 233, letter II; p. 239, letter III).[11]

The Federal Farmer's emphasis on internal police powers is explicable in terms of an understanding of republican government as a government based on principles of private liberty. He considers such a government most conducive to the happiness of the overwhelming majority of the people. Speaking as a genus-liberal, he insists that:

> In free governments the people, or their representatives, make the laws; their execution is principally the effect of voluntary consent and aid; the people respect the magistrate, follow their private pursuits, and enjoy the fruits of their labour with very small deductions for the public use. The body of the people must evidently prefer the latter species of government; and it can be only those few who may be well paid for the part they take in enforcing despotism, that can, for a moment, prefer the former [despotic government]. (FF, p. 264, letter VII)

Indeed, the Federal Farmer's favorite formula for good government is "frugal, free and mild government." In response to the Federalist argument that the critical condition of the country necessitates, almost at any cost, strengthening the federal government, he asserts that "the evils we sustain, merely on account of the defects of the confederation, are but as a feather in the balance against a mountain, compared with those which would, infallibly, be the result of the loss of general liberty, and that happiness men enjoy under a frugal, free and mild government" (FF, p. 334, letter XVII).[12] His own preference is for a government that is limited in nature and *laissez faire* in policy. In this largely negative sense of good government, it is obvious that "a great proportion of social happiness depends on the . . . internal police" (FF, p. 339, letter XVII).

Given the psychological chain he has forged between small-scale government, popular confidence, and voluntary obedience, the Federal Farmer can also justifiably claim that "the detail administration of affairs, in this mixed republics [*sic*], depends principally on the local governments; and the people would be wretched without them" (FF, p. 339, letter XVII).[13] In a consolidated system, where the large-scale federal government exercises internal police powers, everything is reversed. The federal government will be expensive, despotic, and coercive. It will not secure "the freedom and happiness of this people" (FF, p. 229, letter I).

The Federal Farmer, however, does not leave size *per se* as the crucial societal variable. He soon moves on to a richer formulation of the small-republic argument which is informed by the idea of the public happiness or good as something more positive than the absence of governmental restraints on the private liberty of individual citizens and the governmental protection of that liberty from the encroachments of other citizens. The public good is, in addition, the harmony of the more concrete interests of individual citizens or groups of citizens, and a republican government is one which pursues such a good. This second, teleological sense of republican government again points to the states as the American jurisdictions scaled to a republican politics.

Recalling the Federal Farmer's initial statement of the small-republic argument, we can see that it assumes the lack of a public good in the union as a whole: "One government and general legislation alone never can extend equal benefits to all parts of the United States." It also assumes that the realization of a public good depends on social homogeneity: "Different laws, customs, and opinions exist in the different states, which by a uniform system of laws would be unreasonably invaded" (FF, p. 230, letter I). Ironically, the Federal Farmer alleges that it was the great difficulties the Constitutional Convention experienced in reconciling the many conflicts of interest it faced which convinced him that America was too diverse a nation to support a consolidated system of government based on republican principles.

> There were various interests in the convention, to be reconciled, especially of large and small states; of carrying and non-carrying states; and of states more and states less democratic—vast labour and attention were by the convention bestowed on the organization of the parts of the constitution offered; still it is acknowledged there are many things radically wrong in the essential parts of this constitution—but it is said that these are the result of our situation: On a full examination of the subject, I believe it; but what do the laborious inquiries and determinations of the convention prove? If they prove any thing, they prove that we cannot consolidate the states on proper principles . . . (FF, p. 237, letter III)[14]

For the Federal Farmer, an American extended republic implies an incoherence of interests. The federal government can only pursue some narrow interests at the expense of others, which is reason enough for the detailed administration of affairs to remain with the state and local governments.[15]

The presumption of interstate diversity is intrastate homogeneity. The Federal Farmer treats the states as units of interest, in implicit contrast to the union. This contrast is the principal reason the state governments cannot serve as models for the federal government.

> I have often lately heard it observed, that it will do very well for a people to . . . chuse [*sic*], in a certain manner, a first magistrate, a given number of senators and representatives, and let them have all power to do as they please. This doctrine, however it may do for a small republic, as Connecticut . . . can never be admitted in an extensive country . . . (FF, p. 274, letter VIII)[16]

The American diversity also explains why the federal government cannot be a unitary one like the English government, which is exemplary in so many other respects.

> A great empire contains the amities and animosities of a world within itself. We are not like the people of England, one people compactly settled on a small island, with a great city filled with frugal merchants, serving as a common centre of liberty and union . . . (FF, p. 274, letter VIII)

These comparisons suggest yet another meaning of republican government. Their point is not just that the American union, unlike Connecticut and England, is too diverse to support a coherence of interests but that it is also too diverse to support a full and equal representation. The Federal Farmer considers such a representation a necessary (although, again, not a sufficient) condition of republican government. He maintains that the federal government cannot be a republican government in this third, and least consensual, sense of a strongly democratic government, any more than it can be in the sense of a government which is based on principles of private liberty or of a government which pursues an overarching public good.[17]

By a full and equal representation, the Federal Farmer means one which mirrors the citizen body. It "possesses the same interests, feelings, opinions, and views the people themselves would were they all assembled." To achieve such a representation, the electoral process "should be so regulated, that every order of men in the community, according to the common course of elections, can have a share in" the legislature. The representation, though, "must be considerably numerous" to permit "professional men, merchants, traders, farmers, mechanics, etc. to bring a just proportion of their best informed men respectively into the legislature" (FF, p. 230, letter II).

This passage defines the Federal Farmer's normative theory of representation. He argues that all (sizable?) orders or groups of citizens should be physically represented in the legislature.[18] He also argues that they should be represented by their "best informed" members. His theory is not strictly democratic; it embodies the traditional republican idea of a natural aristocracy. But his theory is strongly democratic, both because he denigrates the need for brilliant talents in the legislature and because he stresses the need for a sympathetic legislature (see FF, pp. 265, 268–69, letter VII; p. 292, letter XI; p. 298, letter XII).

The Federal Farmer justifies a full and equal representation in terms of the public good. Different groups of citizens must have their "own" representatives in government because members of one group poorly represent another. By itself, the right of a group of citizens to vote for government representatives is hopelessly inadequate to preventing other groups from utilizing the government to oppress its interests. He turns to Montesquieu and Cesare Beccaria, a contemporary Italian utilitarian, for confirmation of this thesis.

> It is extremely clear that these writers had in view the several orders of men in society, which we call aristocratical, democratical, merchantile [*sic*], mechanic, &c. and perceived the efforts they are constantly, from interests and ambitious views, disposed to make to elevate themselves and oppress others. Each order must have a share in the business of legislation actually and efficiently. It is deceiving a people to tell them they are electors, and can chuse [*sic*] their legislators, if they cannot, in the nature of things, chuse men from among themselves, and genuinely like themselves. (FF, p. 266, letter VII)

In this third sense of the public good as the absence of group oppression, the Federal Farmer also expects the federal government to be deficient, and necessarily so. It cannot fail to pursue policies which oppress certain groups of citizens because a full and equal representation is not feasible on the federal level. In view of the size constraints on deliberative bodies, he insists that it is impossible to properly draw a single representation from a nation as large and diverse as America (see FF, p. 230, letter II; p. 236, letter III; p. 298, letter XII; p. 338, letter XVII).[19]

As always, there is a flip side. The Federal Farmer contends that the state governments not only can but do enjoy equal because full representations (see FF, pp. 232–33, letter II; p. 242, letter III; p. 300, letter XII; pp. 338–39, letter XVII).[20]

Now, it is true that the Federal Farmer believes the federal representation will be unequal in a nonrandom way. He is convinced that it will be dominated by upper-status occupational groups, which he sometimes lumps together as "the aristocratic order." He charges that the federal government will "possess the soul of aristocracy" because even the House "will have but very little democracy in it" (FF, p. 235, letter III; p. 276, letter IX).[21] He is also convinced that while the unequal federal representation will be socially aristocratic, the equal state representations will remain socially democratic. His most strident complaint against the Constitution is that it will transfer power from the many, in the state governments, to the few, in the federal government. He claims that all Americans suspect the general tendency of the plan "is to collect the powers of government, now in the body of the people in reality, and to place them in the higher orders and fewer hands" (FF, pp. 276–77, letter IX).

On one level, however, the Federal Farmer's critique is a dispassionate, scientific critique. A sufficiently democratic federal representation was not available to the framers of the Constitution, whatever their motives might have been. This is not to deny that he often casts their work in harsh, conspiratorial terms, as part of an aristocratic plot against the American democracy (see FF, pp. 225–26, letter I; p. 253, letter V; pp. 258–59, letter VI; pp. 275–76, letter IX; p. 285, letter X).[22] Still, it is not the aristocratically biased federal representation *per se* which he finds so worrisome. After all, that "error" was, to a large extent, unavoidable. What he finds really worrisome is the presence of such a representation in a government which will be so powerful. And although the consolidationist error was avoidable, it was one even the best-intentioned men were liable to commit. The Federal Farmer himself had once committed it, for he, as they, was swayed by the exigencies of the union (see FF, p. 229, letter I; p. 276, letter IX).

The Federal Farmer's prediction that the federal representation will suffer from a strong aristocratic bias is situated within an empirical theory of representation in which size, once more, is the most important independent variable. He asserts a general law of voting behavior that probabilistically associates size of electoral district with outcome. The larger the district, the more likely a member of the aristocratic order is to be elected. Social status translates into name recognition, which is an irresistible cue to voters in large districts.

> A man that is known among a few thousands of people, may be quite unknown among thirty or forty thousand. On the whole, it appears to me to be almost a self-evident position, that when we call on thirty or forty thousand inhabitants to unite in giving their votes for one man, it will be uniformly impracticable for them to unite in any men, except those few who have became [*sic*] eminent for their civil or military rank, or their popular legal abilities . . . (FF, p. 276, letter IX)[23]

According to his own "conspicuous-talents" theory of representation, the Federal Farmer admits that any practical-sized representation, even in small nations or states, will suffer from an aristocratic bias. Yet, he favors some aristocratic bias. In small jurisdictions—assuming that the representation has not been artificially restricted—his empirical and normative theories intersect on "respectable" democrats and aristocrats in rough proportion to their numbers in the general population. These two social types are distinguished by whether they "hold not a splendid, but a respectable rank in private concerns" (FF, pp. 275–76, letter IX).[24] A socially mixed representation, though, is not attainable in large nations, unless they are federated. The Federal Farmer is convinced that the federal representation, unlike the state representations, will far surpass the limits of desirable bias. Due to its unavoidably large electoral districts, the House of Representatives will be staffed almost exclusively by high-toned aristocrats. The only ostensibly democratic part of the new federal government will then also be too elevated from the mass of the citizen body to be sympathetic with it (see FF, p. 269, letter VII).

The Federal Farmer establishes the manifest undesirability of such an unequal representation through emendations to two of his fundamental beliefs, that members of one group make poor representatives of another and that the true end of government is the public good or happiness. He claims that the few (aristocrats) make especially poor representatives of the many (democrats) and that once in office the few will systematically pursue an "unequal" happiness.

History confirms these claims. Rome serves as the Federal Farmer's counterexample of what happens when there is an extremely small, aristocratically biased representation.

> Among all the tribunes the people chose for several centuries, they had scarcely five real friends to their interests. These tribunes lived, felt and saw, not like the people, but like the great patrician families, like senators and great officers of state, to get into which it was evident, by their conduct, was their sole object. These tribunes often talked about the rights and prerogatives of the people, and that was all for they never attempted to establish equal liberty: so far from establishing the rights of the people, they suffered the senate, to the exclusion of the people, to engross the powers of taxation . . . (FF, p. 273, letter VIII)[25]

The Federal Farmer forecasts the same grim future for America under the Constitution. He fully expects the Constitution "to bestow on the former [the higher orders] the height of power and happiness, and to reduce the latter [the body of the people] to weakness, insignificance, and misery" (FF, p. 277, letter IX). The goal should, instead, be to form governments calculated to promote "equal happiness and advantages" (FF, p. 228, letter I).[26]

The last stage in this process of declension—coming on the heels of unequal representation, unequal liberty, and unequal advantages—is unequal social conditions. If the Constitution is put into effect unamended, the Federal Farmer firmly believes that it will destroy America's "happy condition" of unparalleled social equality within "fifty years" (FF, pp. 251–52, letter V). However, the first stage in this bleak, typically republican, scenario is consolidation. For the Federal Farmer, everything else follows as a matter of scientific prediction from the consolidated nature of the Constitution. Because of American geography, movement away from federal principles is movement away from republican equality. This traditional inference stands as his central critique of the proposed plan. As one of the "true friends of a federal republic," he will endorse amendments that "are well defined, and well calculated, not only to prevent our system of government moving further from republican principles and equality, but to bring it back nearer to them" (FF, p. 258, letter VI).

Nontraditional Views: America as a National Republic

Paradoxically, the same equality of conditions which is so threatened by complete consolidation allows for the possibility of partial consolidation. It provides the key to an alternate description of America as a national republic. In his social analysis, the Federal Farmer seems torn between demonstrating that: (1) as implied by the traditional small-republic argument, America is too socially diverse for the federal government to secure republican equality; and (2) because of American exceptionalism, there is a basic equality of conditions that permits the federal government to secure a tolerable degree of republican equality.

The tensions in the Federal Farmer's social analysis appear as the tensions between traditionalism and nationalism. The extent to which his perceptions of American conditions dictated his normative position, or the converse, will always be an open question. To some extent he seems to have embraced nationalism because of those perceptions and to some extent in spite of them. He even seems to have done so in self-conscious opposition to traditional republican principles. Still, he did not stray very far in a nationalist direction precisely because a powerful blend of traditional republican principles and social observations pushed him in the other direction, so much so that he understood republican principles as federal-republican principles. His fundamental federal republicanism is all the less surprising once his intellectual caution is added to the equation of his political thought.

How self-conscious the Federal Farmer was of the tensions in his political thought is another one of those unanswerable questions. He certainly was not conscious of them in the terms I have just articulated them. For him, the essential point was not the tensions between his theoretical commitments but that they somehow fit together. And they did fit together behind his preferred plan of partial consolidation.

Accordingly, the Federal Farmer professes to believe in a number of social "facts" which support the possibility of an American national republic. These facts bolster his preference for partial consolidation and serve to partially assimilate his nationalism to his commitments to traditionalism and social realism.

The Federal Farmer analyzes American society in several different ways, most significantly into two (mega-)orders of aristocrats and democrats and into its multifarious occupational groupings. He confusingly slides between these two modes of social analysis. Nonetheless, a consistent message of homogeneity amidst diversity, even on a national scale, runs through both modes. With respect to his dualistic model of American society, the Federal Farmer suggests that the most conspicuous fact about America is how it does not fit that model. From a trans-Atlantic perspective, all Americans belong to the democratic order (see FF, p. 235, letter III; p. 267, letter VII).[27] Any two societies obviously can be analogized in terms of their hierarchical character and that character can, in turn, be analyzed in terms of social orders. The Federal Farmer's usages of orders in this sense, though, point more toward a feared future than the propitious present. His *idée fixe* is that America not become as hierarchical as European societies (see FF, pp. 251–52, letter V; p. 266, letter VII). Similarly, his more demographic analysis of American society by occupations unearths homogeneity in the form of a vast majority of yeoman farmers (see FF, p. 242, letter III; p. 268, letter VII). To the degree that occupational diversity protrudes, as between agrarian and commercial classes, he suggests that any conflicts of interest between them are more apparent than real. He reminds them that "they, in fact, mutually depend on each other" (FF, p. 267, letter VII).[28]

We, however, can follow the Federal Farmer even farther down this national-republican path of social analysis into the domain of shared beliefs. Although he does not emphasize the softer ideological aspects of social homogeneity (or diversity) as much as the more concrete economic ones, his response to the Federalist argument against a bill of rights does assert a national liberal consensus. He notes that one of the Federalists' avowed reasons for not including a bill of rights in the Constitution is the interstate diversity of opinion and practice which exists in the area of rights. He is quick to show how that argument undercuts their own case for consolidation. Yet, the Federal Farmer seems to undercut his own anticonsolidationist position in concluding that this is one area where not that much interstate diversity exists because American conceptions of rights are part of the nation's common, English heritage (see FF, p. 232, letter II).[29]

The Federal Farmer regularly returns to this common political heritage in the course of his letters, as if to stress how at the deepest levels of belief (and, we today would add, ethnicity) Americans are one (see FF, p. 247, letter IV; p. 321, letter XV; pp. 324–25, 328, letter XVI). He also refers to several other measures of ideological homogeneity: Amidst the polyglot of Christian sects, there is the unity of a Christian America; even if republican principles are embattled in some circles, the overwhelming majority remains firmly attached to a "simple and frugal republicanism" (see FF, p. 249, letter III; p. 348, letter XVIII).

Nor, does the Federal Farmer completely despair of the capacities of the federal government to honor republican principles. America's exceptional equality of conditions presents the possibility of the federal government pursuing a public, and a public as equal, happiness. That possibility increases to the extent that the federal representation can be made "tolerably equal" by immediately doubling it (see FF, p. 277, letter IX; p. 298, letter XII).[30] The further implication is that such a federal government will be able to attract a sufficient degree of popular confidence so that it will not be forced to violate basic principles of private liberty. The Federal Farmer ultimately decides that the federal government can be made republican or democratic enough to receive many of the powers the Constitution would grant it.

While the Federal Farmer offers a series of empirical observations which at least implicitly link American nationalism to republicanism, it is hardly astonishing that explicitly he treats nationalist and republican principles not as mutually supporting but as antagonistic. It is hardly astonishing because on the other side of his intellectual problematic is a tradition of ideas that associates republican societies and governments with small nations or federations of small states. As he himself elaborates that tradition, it means that the federal government cannot directly promote the primary liberal ends of liberty and happiness; that it cannot be sufficiently democratic with any practical-sized representation to be a republican government; that it cannot enjoy enough popular confidence to freely execute its own laws; that the diversity of opinions and interests in the union cannot be managed by one government; that rather than America's fortuitous equality of conditions permitting a transfer of power to the federal level, those conditions counsel against such a transfer of power; and, finally, that the state governments must always be more powerful than the federal government in any system of republican government. For the Federal Farmer, theoretical coherence lay in the extent to which he remained faithful to traditional federal-republican principles.

Significantly, the Federal Farmer closes his letters by declaring that the Constitution is "in many parts, an unnecessary and unadviseable [*sic*] departure from true republican and federal principles" (FF, p. 349, letter XVIII). This passage underscores my point and counterpoint on the Federal Farmer's intellectual dilemma. He can only view nationalism in terms of departures from traditional federal-republican principles; yet he does consider some departures necessary and advisable on nationalistic grounds.

The Federal Farmer marshals a variety of nationalistic formulas to justify limited departures from traditional federal-republican principles. For example, those principles mandate no standing armies, especially on the federal level. He, however, accedes to the establishment of a small national army with the gloss that "after all the precautions we can take, without fettering the union too much, we must give a large accumulation of powers to it, in these [military] and other respects" (FF, p. 343, letter XVIII). Those principles also prescribe a strict division of the sword and the purse between two distinct bodies or levels of government. But, again, this dogma must be breached for the sake of the union. He recommends combining some sword and purse powers in the federal government because "[w]e must yield to circumstances, and depart something from this plan [a strict division of the sword and the purse], and strike out a new medium, so as to give efficacy to the whole system, [and] supply the wants of the union" (FF, p. 282, letter X).[31]

Perhaps, the Federal Farmer just believes that the old federal system of government is not functioning very well or that America's geopolitical situation demands a more powerful central government. Although it is true that he believes the confederation is not functioning very well, his criticisms are extremely temperate compared to those emanating from the other side of the debate. He does not think traditional federal-republican principles have worked that badly in the past, either in America or in Europe (see FF, pp. 224–26, letter I; p. 260, letter VI; p. 277, letter IX; pp. 332–34, letter XVII). Nor, does he think Americans need worry about foreign dangers. He only alludes to the possibility to dismiss it (see FF, p. 225, letter I; p. 349, letter XVIII). He considers the Articles of Confederation defective not simply, or even primarily, on practical grounds. The Articles are also inadequate to "the exigencies of the union" and "the sovereignty of the nation" (FF, pp. 226, 229, letter I). It is mainly on these nationalistic grounds that he judges the Constitution the superior plan. He seems deeply committed to the union, as becomes apparent whenever he takes up the vexatious question of how to divide the sword and purse powers between the two levels of government.

> But it is asked how shall we remedy the evil [of the combination of sword and purse powers in the federal government], so as to complete and perpetuate the temple of equal laws and equal liberty? Perhaps we never can do it. Possibly we never may be able to do it in this immense country, under any one system of laws however modified; nevertheless, at present, I think the experiment worth a making. I feel an aversion to the disunion of the states, and to separate confederacies; the states have fought and bled in a common cause, and great dangers too may attend these confederacies. (FF, p. 281, letter IX)

These "great dangers" indicate the practical advantages of the union in deterring civil wars. The Federal Farmer earlier had asserted both that "the greatest political evils that can befal [*sic*] us, are discords and civil wars" and that the union is one of "the greatest blessings" (FF, pp. 257–58, letter VI). But he regards the union as a blessing not only because civil wars are evil. Civil wars are evil, in part, because the union is a blessing.

The Federal Farmer is genuinely captivated by the American mission, a mission which is inseparable from the union. In the passage quoted above, he defines the mission as an "experiment" in perpetuating "equal laws and equal liberty" in an "immense country."[32] For him, as for his opponents, it is an experiment in republican government on a grand scale. It would not be such an experiment if the union did not, in an important sense, meet traditional republican specifications. Due to the change in scale, though, the experiment also requires a relaxation of those specifications. The Federal Farmer participated in both intellectual moves. He did take considerable pride in the experiment and he was powerfully influenced by American nationalism.[33]

Still, the Federal Farmer separated himself from his opponents through his unwillingness to go very far in a nontraditional, national-republican direction. In practice, as in thought, he sought to cautiously limit the great American experiment so that it did not lose its essential republican character. He wanted it to proceed in gradual steps, as befitted a true, scientifically controlled experiment. Federal powers should be increased. They, however, should not be increased prematurely, before the federal government has been fitted for them with a more equal representation (see FF, pp. 240–41, 243, letter III).[34] He was a cautious man. His alternative constitution is incomprehensible without taking that factor into account. He hesitated before the novelty of an extended republic and the national government it would impose on the American states.

The Federal Farmer's plan of partial consolidation reflects each of the antipodal tendencies in his political thought. It merges his theoretical caution with his experimentalism; his traditionalism with his criticism of tradition; and his state-centric social analysis with his nationalism. His plan actually represents a substantial retreat from his initial vision of complete consolidation.

"Partly National, Mostly Federal"

The Federal Farmer is conscious of his plan as a nontraditional one. He positions it between the Constitution and the Articles of Confederation. Where he sees the first as a widely experimental plan of complete consolidation, he sees the second as an overly traditional federal plan. His alternative, though, is a mostly traditional, mostly federal plan. Structurally, it is closer to the Articles than it is to the Constitution, in spite of his preference for the latter on nationalistic grounds.[35]

Admittedly, the Federal Farmer sometimes redefines federalism so that it means partial consolidation. He does this when analyzing the Anti-Federalist and Federalist parties (in a manner which nicely captures the ambiguity of their party labels).

> Some of the advocates are only pretended federalists; in fact they wish for an abolition of the state governments. Some of them I believe to be honest federalists, who wish to preserve substantially the state governments united under an efficient federal head . . . Some of the opposers also are only pretended federalists, who want no federal government, or one merely advisory. Some of them are the true federalists, their object, perhaps, more clearly seen, is the same with that of the honest federalists . . . We might as well call the advocates and opposers tories and whigs, or any thing else, as federalists and anti-federalists. (FF, p. 258, letter VI)[36]

For the most part, however, he acknowledges that his preferred plan departs from traditional federal principles (see FF, p. 229, letter I; pp. 332, 337, letter XVII; pp. 343, 349, letter XVIII).[37]

According to the Federal Farmer, traditional federal principles are primarily defined by different modes of exercising powers as between the federal and state governments. He claims that "the quantity of power the union must possess is one thing, the mode of exercising the powers given, is quite a different consideration; and it is the mode of exercising them, that makes one of the essential distinctions between one entire or consolidated government, and a federal republic" (FF, p. 331, letter XVII). He does not argue the point of whether a federal system requires equal state suffrage or the original consent of the state governments, nor does he fasten on a particular division of powers between the federal and state governments. He opens his letters by confessing that he is "not disposed to unreasonably contend about forms" (FF, p. 224, letter I).[38] This is not to deny that one of his constant refrains in attacking the Constitution is the unprecedented powers it grants the federal government (see FF, p. 229, letter I; p. 233, letter II; p. 239, letter III; p. 334, letter XVII; pp. 339–40, letter XVIII).[39] What he means by powers in these contexts, though, is typically ambiguous. He may either mean the general authority over certain objects of government (such as defense) or the more specific authority to exercise that "power" in certain ways (such as to raise armies). One indication that his primary reference is to the second sense of powers is that there is no object of government the Constitution places under federal supervision which he would return to the states, except perhaps for the relatively minor power over bankruptcy regulations.[40]

The particular way in which the Federal Farmer understands the differences between the Articles and the Constitution supports this conclusion. The commerce and currency powers are the Constitution's most significant additions to federal powers in the sense of objects. The Federal Farmer, however, unhesitantly accepts those additions (see FF, p. 229, letter I; p. 239, letter III; p. 260, letter VI).[41] On the two most important objects of government—the sword and the purse—the Articles and the Constitution are "only" distinguished by the modes of power they confer upon the federal government. Yet, it is this difference which, according to the Federal Farmer, defines the Articles as a federal plan, in essential harmony with traditional federal principles, and the Constitution as a consolidated plan, in massive violation of those principles. He insists that "a federal head never was formed, that possessed half the powers which it can carry into full effect, altogether independently of the state or local governments, as the one, the convention has proposed, will possess" (FF, p. 339, letter XVIII).[42]

The Federal Farmer, nevertheless, is critical of the Articles, a judgment which suggests the inadequacy not only of traditional federal principles but of any strict division of modes of exercising powers between the federal and state governments, even on the sword and the purse. In his opinion, the optimal system of government must be at least partly nontraditional and, by the same token, partly nonfederal.

The Netherlands, more than any other past or present federation, informs the Federal Farmer's understanding of traditional federal principles. He mentions it as a possible precedent in three respects: the distinction between external and internal taxation, requisition systems, and state checks on the federal government. In each instance, though, he would deviate from Dutch practice in a way which strengthens the federal government. He would not confine the federal government to external taxation, compel it to raise other revenues through requisitions on the state governments, or subject its laws to single-state vetoes (see FF, p. 239, letter III; pp. 337–38, letter XVII).[43]

The Federal Farmer's references to other political writers confirm the impression that he favors new, less rigid conceptions of federalism.

> In a federal system we must not only balance the parts of the same government, as that of the state, or that of the union; but we must find a balancing influence between the general and local governments—the latter is what men or writers have but very little or imperfectly considered. (FF, p. 260, letter VI)

He later promises to "pursue a series of checks which writers have not often noticed" (FF, p. 283, letter X).

This "series of checks" constitutes a self-consciously new idea of a federal balance. Analogous to the idea of a balance of powers between different departments of government, the Federal Farmer's idea of a federal balance between different levels of government permits a tremendous amount of flexibility in distributing powers. The *sine qua non* in each case is to maintain the balance.

The Federal Farmer presents a confusing array of proposals on how various powers should be divided between the federal and state governments. What gives these proposals coherence is his idea of a federal balance.

> Here again I must premise, that a federal republic is a compound system, made up of constituent parts, each essential to the whole: we must then expect the real friends of such a system will always be very anxious for the security and preservation of each part, and to this end, that each constitutionally possess its natural portion of power and influence—and that it constantly be an object of concern to them, to see one part armed at all points by the constitution . . . and the others left constitutionally defenceless. (FF, p. 341, letter XVIII)[44]

In this light, the Federal Farmer views both the Constitution and the Articles as unbalanced. The Constitution distributes the powers of government in such a manner as to leave the state governments defenseless. The Articles, however, can legitimately be criticized on the opposite grounds, for exposing the federal government to the charity of the state governments (see FF, p. 226, letter I; p. 260, letter VI; pp. 282–83, letter X; p. 339, letter XVII; p. 341, letter XVIII).[45]

The Federal Farmer is unsure of exactly how to institutionalize a federal balance. Evidently, it is an idea which is open to experimentation.

He weighs three options: (1) the federal government indirectly exercises a certain power through the state governments, but it possesses provisional powers to act directly on individual citizens in the case of state noncompliance;[46] (2) it directly exercises a certain power, but that power is narrowly granted; and (3) it directly exercises a certain power but only through restrictive legislative procedures involving congressional supermajorities and collective state vetoes.

Reflecting an experimental mode of thought, the Federal Farmer's proposals on how the decisive authority over the sword and the purse should be divided between the federal and state governments include all three options. As we have seen, he heavily censures the Constitution for the way in which it deposits those powers, almost without limit, in the federal government. Still, he would distribute them in a way which is more favorable to the federal government than is the case in the Articles.

With regard to the defense powers, the Federal Farmer recommends: (1) retaining a requisition system for raising federal armies, supplemented by federal authority to raise its "own" army if the state governments fail to fill their quotas; (2) a one-year limit on military appropriations, and a 2,000-man limit on any "standing" army in times of peace and 12,000-man limit in times of war; and (3) federal laws authorizing the above to require the approval of a two-thirds or three-fourths congressional majority and/or subjecting such laws to a veto by state legislatures which together represent a majority of the American people (see FF, p. 338, letter XVII; p. 343, letter XVIII).[47]

The Federal Farmer advocates the same coercive requisition system and extraordinary legislative procedures for federal taxes (see FF, pp. 337–38, letter XVII). As far as grants of authority, he would deny the federal government concurrent taxing powers (except in cases of state noncompliance) because he believes they cannot be balanced.

> Besides, to lay and collect internal taxes, in this extensive country, must require a great number of congressional ordinances, immediately operating upon the body of the people; these must continually interfere with the state laws, and thereby produce disorder and general dissatisfaction, till the one system of laws or the other, operating upon the same subjects, shall be abolished. These ordinances alone . . . will probably soon defeat the operations of the state laws and governments. (FF, pp. 239–40, letter III)[48]

Yet, the Federal Farmer's balance clearly leans in the opposite direction, toward the state governments. It defines an alternative constitution which upholds the core of traditional federal principles and only mildly deviates from prior American (and Dutch) practices. He offers his own experiments as ones carefully designed to enhance those principles, not to recklessly discard them—which is how he understands the Constitution. Indeed, he develops what looks very much like a methodological critique of the proposed plan, as if its framers' ultimate goals and his own were the same and they merely differed in their political audacity.

The Federal Farmer's fundamental methodological assumption is that "good government is generally the result of experience and gradual improvements" (FF, p. 260, letter VI). He castigates "a rage for change and novelty in politics" (FF, p. 329, letter XVI). He portrays the Constitution as a violation of that assumption and as a product of that trend. It rests "all on logical inference, totally inconsistent with experience and sound political reasoning" (FF, p. 334, letter XVII).[49] He repeatedly uses such adjectives as "novel," "new," and "unprecedented" to characterize various features of the plan (see FF, p. 233, letter II; p. 253, letter V; p. 280, letter IX; p. 329, letter XVI; pp. 336–38, letter XVII; pp. 339, 344, letter XVIII).

In an extremely revealing passage in his last letter, the Federal Farmer explains how Americans should proceed in empowering a new federal government.

> I think there is a safe and proper medium pointed out by experience, by reason, and facts. When we have organized the government, we ought to give power to the union, so far only as experience and present circumstances shall direct, with a reasonable regard to time to come. Should future circumstances, contrary to our expectations, require that further powers be transferred to the union, we can do it far more easily, than get back those we may now imprudently give. The system proposed is untried: candid advocates and opposers admit, that it is, in a degree, a mere experiment, and that its organization is weak and imperfect; surely then, the safe ground is cautiously to vest power in it, and when we are sure we have given enough for ordinary exigencies, to be extremely careful how we delegate powers, which, in common cases, must necessarily be useless or abused, and of very uncertain effect in uncommon ones. (FF, p. 336, letter XVII)[50]

The Federal Farmer implicitly sets up a battle of logics with his adversaries. His is sociological; theirs is philosophical. His is inductive; theirs is deductive. His is pragmatic; theirs is idealistic. Most of all, he counterposes his caution to their rashness. But behind, or, rather, outside, his political logic was a normative commitment to a federal, and a federal because republican, tradition of political ideas. What was paramount for him was not to put the American republic, as defined by that tradition, at risk. From our perspective, the Federal Farmer seems to have abruptly halted his twin movement in social perception and government structure toward national republicanism at the point where he thought it threatened the future viability of the American republic—really, the American republics. To a cautious man, there are always formidable dangers in innovation, even if he is proud of his own innovations. Fascinating reformulations of traditional ideas can lead to tragic abandonments of salutary beliefs; architectonically flawless institutions can become detached from benevolent social conditions. To the Federal Farmer, the Constitution embodied those dangers.

4

The "Nationalist" Papers

Publius does not concede the mantle of logical consistency to his opponents. On the contrary, he repeatedly scolds them for their inconsistencies. "Pragmatic" would probably be the last word he would use to describe the Federal Farmer or any other Anti-Federalist.[1]

The argument of the first three "heads" (papers 1–36) of *The Federalist Papers* is eminently logical. That argument is: (1) the union is necessary to secure the political happiness of the American people; (2) an energetic federal government is necessary to preserve the union; (3) the Constitution provides for an energetic federal government; (4) the American people, therefore, should support the Constitution.[2]

But, of course, a logical argument is only as truthful as its premises. The Federal Farmer questioned the first and second premises and, thus, denied the conclusion. He claimed that even though a more energetic federal government was necessary to preserve the union, it did not have to be as energetic as the one proposed. He also claimed that even though the union was a desirable political arrangement, it was not essential to the political happiness of the American people. He thought that the individual states were more intimately connected to the ends of any political society and that a too energetic federal government imperiled those ends by undermining the existence of the states as quasi-independent societies.

The Federal Farmer's position was not really illogical. Given his premises, it was not contradictory for him to favor continued union but not the Constitution; just as, given his premises, Publius could justifiably insist that such a position was contradictory. The two founders held distinct sets of perceptions about the causes and gravity of the prevailing crisis, the lessons of history, and the international situation of their country.

While the Federal Farmer and Publius disagreed about the best geopolitical means to the ends of politics under American conditions, they did not disagree about the nature of those ends. Both statesmen were liberals who defined the primary political ends as liberty and happiness. Presumptively, however, Publius was a pluralist liberal insofar as he viewed the larger union as the best means to liberty and happiness under American conditions; the Federal Farmer, a liberal republican insofar as he viewed the smaller states in those terms.[3]

The validity of the second presumption was *mostly* accepted in the preceding chapter. The validity of the first will mostly be *rejected* in this chapter.

One of the major themes of *The Federalist Papers* is to portray the union as a republic, and as a better republic than the individual states. Publius's extended-republic argument challenges traditional republican assumptions. He, nevertheless, characterizes it as a reformulation, not a renunciation, of those assumptions. He conceives of himself as an innovative republican theorist who has intellectually legitimated an expansion of the sphere of republican government. He also conceives of himself as a national republican who has demonstrated that in America republican government will work better on a national than on a state scale. In these terms, he could unhesitantly proclaim his nationalism; his federalism could only be rhetorical.[4]

Conversely, federalism is the Federal Farmer's touchstone position. It is his nationalism which seems not so much rhetorical as muted. Both of these attitudes follow from a relatively straightforward application of traditional republican assumptions to American conditions. The Federal Farmer opposes his largely traditional federal republicanism to Publius's more novel national republicanism.

However, this use of "isms" to categorize the differences between the Federal Farmer and Publius may only obscure what basically were two different readings of American conditions. Even if we are sure that deeper theoretical differences informed their different readings of American conditions, it proves difficult to articulate the exact nature of those deeper differences without imposing our categories of thought on theirs.

We are fairly confident that Publius held a set of beliefs which was nationalistic in our sense of the term and that those beliefs provided a normative stance which was partially independent of traditional republican beliefs. To invoke a well-tested formula, he valued the union as both a means and an end. Still, his differences with the Federal Farmer seem to lie less in a deeper commitment to the union than in the greater extent to which he thought that commitment could be supported on traditional republican grounds.[5]

Yet, it was not the same republicanism which the Federal Farmer and Publius believed worked better on either the state or federal level. We are more confident of this difference. The adaptation of traditional republican beliefs to either the smaller states or larger union could not fail to produce results whose differences went beyond mere differences in scale.

In fitting traditional republican beliefs to the American states, the Federal Farmer could rely on the small-republic argument and he could count on a *prima facie* persuasiveness to his efforts. It was Publius's efforts which required a tremendous amount of pushing and shoving to fit the republican tradition to the union in such a way that the final product appeared realistic yet still sufficiently republican. That tradition, though, could not be "nationalized" in a credible manner without being diluted. If he had been prone to analyze the ratification debate purely in intellectual terms, the Federal Farmer would undoubtedly have taken this perspective on his rival's position; not, or not only, that it embellished their common tradition of political ideas and legitimated an emerging ethos of American nationality to which he also was very sympathetic. Publius's labors, nonetheless, were remarkably successful. In retrospect, his synthesis of traditional political ideas, sociological observations, and nationalist sentiments remained closer to the republican than to the pluralist pole of American liberalism, even if it was more pluralistic than the Federal Farmer's competing synthesis.[6]

At bottom, then, the ratification debate was a dispute over different definitions of both the American republic and republicanism. The Federalists' position was indefensible unless the union was thought of in a certain way and unless republicanism was thought of in a certain way. A mixture of empirical, predictive, normative, and definitional factors was implicated in their controversies with the Anti-Federalists. Even different political methodologies played a significant role. Any adequate interpretation of the debate, and of its individual participants, must take each of these factors into account.

Publius's political thought can best be understood by analyzing the argument of *The Federalist Papers* into two tiers. On the first tier, largely the first three heads, his premise is that America is a nation. His conclusion is that it must possess an essentially nonfederal or national system of government to remain one nation. He, not very subtly, attempts to establish an expansive definition of federalism to include the proposed government. However, it is equally important for him to, *sotto voce*, wrap the proposed government with all the attributes of power and the union with all the attributes of nationhood. His fitting of government to society on this tier is complicated by several predictive and normative factors. He suggests that America, whether it currently is a "full" nation or not, will become more of one in the future provided it stays united; that a national government is necessary to this process; and, therefore, that nationhood is as much a goal as a premise. The ringing nationalist affirmation with which he closes the papers—"a nation, without a national government, is, in my view, an awful spectacle" (Publius, p. 527, paper

85)—reveals that even the sense in which the union "needs" a national government is ambiguous.

The second tier of Publius's political argument involves another fitting, of the republican tradition to the union and its new government. Although the claim that the federal government will be a republican one is peculiarly the province of the fourth head, the claim that the union is a national republic provides a subtext for the entire *Federalist Papers.* This tier also is complicated by contested definitions of the nature of republican governments and societies, normative statements about the path American political development should take, and hopeful predictions about the path it will take.

It is impossible to determine which of the two tiers was more basic to Publius. They appear to have been inextricably blended in his own mind as national republicanism. To a perhaps surprising degree, his papers seem directed to rebutting the traditional presumption against extended republics. The remainder of this chapter will elaborate Publius's bold attempt to dispel an alleged inconsistency on his side of the ratification debate.

The Logic of Union

Publius's disclosure of the republican nature of the union starts in the first head (papers 1–14) with the ends. The argument of these papers is a comparative one. Under American conditions, the ends of any political society are more secure in union than in disunion. By this most fundamental standard of the utility of all political arrangements, the union is a better republic than the individual states; or, more generally, the larger the republic, the better. Completely reversing conventional wisdom, Publius contends that it is small, not large, nations that cannot form viable republics. After asserting the comparative advantages of the union to the states on the ends, the argument that a competent federal government can better protect the ends than equally competent state governments can follows as a matter of course. A partially separate argument is that the federal government will almost inevitably be more competent than the state governments.

At first, Publius's object is to measure the utility of the union. This effort involves the somewhat counterintuitive demonstration that the manner in which the American subcontinent is politically divided will have an enormous impact on the happiness of its residents. Moreover, this whole effort is, in a sense, a red herring. Despite Publius's allegations to the contrary, his opponents are not disunionists who see no value in the union (see Publius, p. 34, paper 1; pp. 37–38, paper 2; p. 97, paper 13; p. 103, paper 14; pp. 151–52, paper 22).

We, nevertheless, must bear in mind Publius's assumption, as later defended in heads two (papers 15–22) and three (papers 23–36), that the establishment of a federal government less energetic than the one proposed will result in disunion. The Anti-Federalists, in denying that assumption and, hence, adequate powers to the federal government, would, in effect, be disunionists. This caveat shows that Publius's charge that the Anti-Federalists want contradictions—union without an energetic federal government—really supersedes his charge that they are disunionists (see Publius, p. 108, paper 15; p. 157, paper 24).

In truth, the rhetorical weight of Publius's argument in the first as well as in the next two heads rests on the assumption that an energetic federal government is necessary to preserve the union, not on the utility of the union. Yet, it is his explication of the utility of the union that illuminates his national republicanism. That explication, furthermore, does carry considerable weight. While Americans might agree that the union is an advantageous political arrangement, they might disagree, or be unclear, about just how advantageous it is and exactly in what ways it is advantageous. In the first head, Publius is trying to persuade such an audience to place greater value in the union, which will then predispose it to assent to the remainder of his argument.

The Liberal Ends

The logic of union in the first head is clear. Papers three through seven argue that the union promotes the public safety. Americans are safer from foreign and domestic dangers living as one nation than they are disunited, living as thirteen, or two or three, nations. In the same comparative terms, papers eight through ten argue that the union protects the private and public liberty of American citizens by decreasing the need for standing armies and the possibility of factional violence. Finally, papers eleven through thirteen argue that the union boosts the economic prosperity of the American people. It supports a flourishing commerce, facilitates the collection of public revenues, and reduces the expensive duplication of government services. (Publius continually emphasizes in these papers that the union must possess its own energetic government for these comparative advantages to be fully realized.)

In the first head, Publius, then, identifies the primary political ends as safety, liberty, and prosperity. Collectively, these three ends define the political happiness of the American people. Publius must believe that this definition will command the immediate assent of his audience since he defends the utility of the union by those ends, but not the ends themselves. They are noncontroversial, liberal ends. The question becomes, how republican or pluralistic is Publius's definition of political happiness. A closer examination of paper ten is critical to answering that question.

Paper Ten

The problematic of this most famous, and most misunderstood, of all the *Federalist Papers* is that popular governments are prey to factions. This tendency comforts the critics of that form of government and disturbs its patrons. Publius calls these two groups, "the adversaries to liberty" and "the friends of public and personal liberty" (Publius, p. 77, paper 10). Both groups focus on the susceptibility of popular governments to factions because they threaten the objects or ends of government. They endanger "the rights of other citizens" and "the permanent and aggregate interests of the community" (Publius, p. 78, paper 10). Restated as a disturbing question, the problematic of paper ten is: Can popular governments be good governments, notwithstanding their universally admitted evil propensity to factions?

An elaboration of the ends which factions threaten is necessary to understand Publius's solution to this problematic. "The rights of other citizens" presumably refers to the same phenomena as the personal liberty its friends fear popular governments cannot adequately protect. Publius provides yet another formulation of this end when he declares that "the first object of government" is the protection of human faculties, including (but not only!) the "different and unequal faculties of acquiring property" (Publius, p. 78, paper 10).[7] He, thus, conceives of this fundamental liberal end of personal or private liberty very broadly, as a general right to self-development, which it is, *prima facie,* unjust for any one to violate and which it is the special duty of government to protect from the violence of factions.

The nature of "the permanent and aggregate interests of the community" is more difficult to determine. Indeed, a crucial part of Publius's solution to the problem of factions concerns just how difficult it is for a community to determine, and systematically pursue, its own permanent and aggregate interests. We can, with some confidence, say that those interests refer to different phenomena, or at least additional phenomena, than the protection of private rights. We also are fairly confident that they refer to the same phenomena as the common, general, or public good, terms Publius uses interchangeably throughout paper ten (see Publius, pp. 77, 79–80, 82, paper 10).[8] These terms seem to define a more concrete, and inevitably more contested, way of thinking about political happiness than through the broad goals of safety, liberty, and prosperity; a way which translates the particularistic interests individual citizens have in, for example, specific tariff rates into policies that somehow find a balance between those interests while serving the long-term interests of the society.

Still, it is not clear that we have exhausted the ends which are at issue in paper ten. The friends of liberty are also friends of public liberty. How does that type of liberty fit into the three-pronged thrust of the argument of paper ten?: How do factions menace public liberty?; how is it an end of government?; and how is its protection a peculiar difficulty for popular governments?

Publius's repetition of the argument of paper ten in paper fifty-one clarifies the status of public liberty. There, he claims, "Justice is the end of government . . . It ever has been and ever will be pursued until it be obtained, or until liberty be lost in the pursuit" (Publius, p. 324, paper 51).[9] The inference is that if popular governments cannot prevent injustice—that is, adequately protect the private rights of individual citizens from hostile factions—then those citizens will themselves abandon that form of government for one which promises to better protect their rights by not permitting them as much freedom to organize factions and otherwise participate in public affairs. In this way, factions endanger both (directly) private liberty and (indirectly) public liberty.

However, Publius is himself determined not to abandon the cause of popular government or public liberty. In paper fifty-one he rejects the possibility of setting up a nonpopular government as a means of suppressing factions, and in paper ten he refuses even to contemplate the possibility of abolishing public liberty because it is clearly a case where the remedy is worse than the disease (see Publius, p. 78, paper 10; pp. 323–24, paper 51).

Publius, therefore, treats public liberty as an end of government, which exists in some tension with the ends of private liberty and the public good. Yet, he will yield none of these ends to the other(s). This tension also indicates how the protection of public liberty is a special difficulty for popular governments. In a nonpopular government, public liberty will always be further restricted when it jeopardizes other ends through its factious spores, and those restrictions will not violate the spirit of the government. The citizens and leaders of a popular government, to remain faithful to its spirit, must resist that easy solution.

Now that Publius has reassured his audience that he will not forsake the cause of popular government, he can narrow his problematic. (Given the character of the ratification debate, this, of course, was something about which he had to reassure his audience.) Because the remedies to the diseases of factions must be found within the forms of popular government, it is only the threat of majority factions which can be fatal. The normal operations of popular governments will (eventually) neutralize minority factions. At this point of the argument in paper ten, Publius restates his own problematic as: "To secure the public good and private rights against the danger of such a [majority] faction, and at the same time to preserve the spirit and form of popular government, is the great object to which our inquiries are directed" (Publius, pp. 80–81, paper 10).[10]

The possibility of popular remedies to popular diseases lies in the fact that popular government is a genus with several species. Publius extols the virtues of extended republics over small republics and democracies. In extended republics, it is less likely that majority factions will form, and it is more likely that the government will be led by enlightened statesmen. These virtues are presented as the negative and positive conditions of protecting private liberty and pursuing the public good in a government that allows its citizens extensive public liberties. The two halves of Publius's solution to the problem of factions will be called the dispersement and refinement principles.[11]

"Pure" democracies offer no cure to the evils of factions. Neither the dispersement nor refinement principle can function in such societies. By nature, they are small and homogeneous, which precludes the inhibitive effects group diversity has on the formation of majority factions. Almost by definition, their nonrepresentative governments are not led by enlightened statesmen, who would single-mindedly pursue the public good irrespective of factional alignments. Publius concludes, "Hence it is that such democracies have ever been spectacles of turbulence and contention; have ever been found incompatible with personal security or the rights of property; and have in general been as short in their lives as they have been violent in their deaths" (Publius, p. 81, paper 10).[12]

Small republics offer more hope. The refinement principle can function in such societies. Sociologically, though, the nature of small republics is similar to pure democracies and precisely for that reason not only does the dispersement principle remain inoperative, but the effects of the refinement principle may well be reversed. What is filtered through the representative process is more likely to be "men of factious tempers, of local prejudices, or of sinister designs" than men "whose wisdom may best discern the true interest of their country and whose patriotism and love of justice will be least likely to sacrifice it to temporary or partial considerations" (Publius, p. 82, paper 10).

Extended republics are unquestionably the superior species of popular government. First, the probability of positive results from the refinement principle increases in proportion to the ratio between representatives and constituents. This ratio is, in turn, directly related to the size of the society. Second, the dispersement principle is now operative. The probability of majority factions is inversely related to the size (and diversity) of the society. It is in the form of large-scale republican governments that popular governments enjoy the greatest potential to be good governments, as defined by the prospects they offer for protecting private liberty and pursuing the public good while maintaining citizenship rights (see Publius, pp. 82–83, paper 10).[13]

Publius closes paper ten by localizing his solution.

> In the extent and proper structure of the Union, therefore, we behold a republican remedy for the diseases most incident to republican government. And according to the degree of pleasure and pride we feel in being republicans ought to be our zeal in cherishing the spirit and supporting the character of federalists. (Publius, p. 84, paper 10)

According to Publius, true republicans are Federalists. But are they true federalists? Are they even true republicans?

It seems indisputable that federalism does not play a major role in the argument of paper ten. The argument, in fact, pivots on a comparison between the federal and state governments that assumes they are analogous in structure in order to establish the (potential) superiority of the federal government. Publius's project in the first three heads is to show how the union is *not* like a federation of small, state republics and how the Constitution is *not* like the Articles of Confederation. The "proper structure of the Union" is essentially not a federal one (see Publius, pp. 83–84, paper 10).[14]

Many scholars insist that the argument of paper ten is equally not a republican one. The validity of this claim is more open to dispute. Paper ten has become a crucible for opposing interpretations of Publius's political thought.

The most frequently cited piece of evidence in pluralist (narrowly liberal) readings of paper ten is Publius's analysis of factions.[15] This analysis presents a vision of a protean society in which factions are endemic; the more, the merrier. It also presents a vision of government as an embattled umpire whose principal task is the regulation of factions. These visions are pluralistic. They are not republican insofar as they suggest an intemperate fascination with factions; a premature surrender of social homogeneity and equality as impractical, if not undesirable, goals; a gleeful ridiculing of civic virtue as an empty appeal; and a facile skepticism that the public good could be anything more than adventitious harmonies amidst a cacophony of group demands on government.

Another indication of Publius's pluralism is how he places private above public liberty in the argument of paper ten, especially in his paper fifty-one rendition of the argument.[16] He implies that the primary duty of government is to protect private rights, not to act as a catalyst for a rich public life. He also seems to value public liberty more as a means for citizens to protect their own rights and interests in the public arena than as an end in itself.

Nonetheless, the evidence for the other, republican interpretation is more compelling.

In the first place, we must be impressed with Publius's determination not to sacrifice public to private liberty. He refers to public liberty, and to the popular governments which nurture it, not just as a means to other ends but as a cause he hopes can be "recommended to the esteem and adoption of mankind" (Publius, p. 81, paper 10). He also considers it "essential to political life," as if political, in distinction to private, life contained its own set of nonfungible demands and rewards (Publius, p. 78, paper 10). Admittedly, he would direct the government to the protection of private rights rather than to the promotion of civic virtue. That emphasis, however, only proves that he was, broadly speaking, a liberal—not necessarily that he was a pluralist liberal.[17]

Second, Publius's vision of government transcends the protection of private rights from factional violence. Although that is the basic action of government, it, in another sense, is merely the enabling condition for a higher pursuit; namely, the pursuit of a public good which is simply defined by neither the protection of private rights nor the balancing of group interests. Publius's "enlightened statesmen" substantiate this more elevated vision of government, for they "adjust these clashing interests and render them all subservient to the public good" (Publius, p. 80, paper 10). In paper ten, he strongly suggests a bifurcated public-policy process fueled by frictions between group interests and a greater good and perplexed by schemes to give the former priority over the latter. He intimates that this greater good has an existence apart from group interests, as if it were an ideal substance, discernible by those of sufficient wisdom to discern it and pursuable by those of sufficient virtue to pursue it. In short, Publius's vision of government incorporates traditional republican ideas of the public good and a natural aristocracy.[18]

Third, Publius's vision of society is not strictly, nor even principally, pluralistic. Although he rejects the traditional republican assumption of social homogeneity and accepts a multiplicity of factions in paper ten, that rejection and acceptance can hardly be called wholehearted. They are colored by his belief that factions are necessary evils in any republican society, no matter how small, homogeneous, or traditional it may be.

The proliferation of factions in a large, rapidly developing republic forestalls the worse, and unnecessary, evil of majority factions, but it certainly does not dispel all the ill-effects of factions. Besides, the very problematic of the paper defines factions negatively, in normative terms. Publius's definition of factions is fully consistent with the traditional republican antipathy toward political parties and factions.[19]

One final point: While it is true that Publius also dismisses the closely related assumption of social equality and mocks Americans who suppose that the equal rights of a republican government will introduce an equal division of property, the inequality he supports can, again, be subsumed under the traditional republican idea of a natural aristocracy. Publius, after all, does associate unequal property with unequal talents (see Publius, pp. 78, 81, paper 10).

In view of the conflicting tendencies in paper ten to disclaim, affirm, modify, and exploit the tensions between traditional republican beliefs, we can best locate Publius at an intermediate position between republicanism and pluralism. Yet, he still appears mostly republican. An analysis of the other *Federalist Papers* confirms this assessment. It also illustrates how the pluralist strand in Publius's political thought was the price of a credible national republicanism, a price he was very willing to pay (though he, of course, would not have understood it precisely in those terms).[20]

Republican Definitions of Union

Misinterpretations of Publius's political thought not only arise from exaggerated readings of paper ten but from an undue focus on that paper at the expense of the other eighty-four papers.[21] In this section of the chapter, I will analyze two different ways he defined the union as a cohesive national republic in the "other" *Federalist Papers*. This rhetorical purpose was not evident in paper ten, where Publius, perhaps more realistically, seemed intent on exploring the benefits of a lack of cohesion. In this respect, his efforts elsewhere are more traditionally republican. They, nevertheless, are not completely successful on those grounds, nor are they exclusively pursued on those grounds.

A Union of Interests

Publius never seems to doubt that the union supports an overarching public good and that it, therefore, is a republic in this traditional sense. In paper ten, he contrasts the great interests of the nation to petty local interests (see Publius, p. 83, paper 10). As a national republican, however, he must show, more concretely, how such national interests exist, something the argument of paper ten seriously calls into question.

Twice, Publius explicitly depicts the interests of the nation as a coherence of the interests of its parts. In the course of discussing the advantages of a strong union in negotiating favorable treaties with other countries, he contends that the federal government "can harmonize, assimilate, and protect the several parts and members, and extend the benefit of its foresight and precautions to each. In the formation of treaties, it will regard the interest of the whole, and the particular interests of the parts as connected with that of the whole" (Publius, pp. 47–48, paper 4). The second occasion also occurs when he is discussing the treaty powers, this time in regard to the improbabilities of the Senate approving treaties which sacrifice the interests of some of the states. He notes that "the government must be a weak one indeed if it should forget that the good of the whole can only be promoted by advancing the good of each of the parts or members which compose the whole" (Publius, p. 395, paper 64).

One way Publius specifies this general claim of a national coherence of interests is to rebut the argument, commonly heard on the other side of the ratification debate, that the two mega-interests of agriculture and commerce are antagonistic. He insists that those two interests are, in fact, harmonious.

> The often-agitated question between agriculture and commerce has from indubitable experience received a decision which has silenced the rivalship that once subsisted between them, and has proved, to the entire satisfaction of their friends, that their interests are intimately blended and interwoven. It has been found in various countries that in proportion as commerce has flourished land has risen in value . . . Could that which procures a freer vent for the produce of the earth, which furnishes new incitements to the cultivators of land, which is the most powerful instrument in increasing the quantity of money in a state . . . which is the faithful handmaid of labor and industry in every shape fail to augment the value of that article [land], which is the prolific parent of far the greatest part of the objects upon which they are exerted? (Publius, pp. 91–92, paper 12)[22]

In particular, Publius claims that tariffs provide a source of government revenues which is relatively painless to both interests, certainly more so than property taxes (see Publius, p. 96, paper 12).[23]

Another way Publius specifies this coherence of interests is in repudiating the Anti-Federalist ideal of a full and equal representation. He contends that insofar as both the agricultural and commercial interests are internally homogeneous, again particularly in relation to taxation, the representation of each of their subinterests is superfluous (in addition to being impractical and undesirable) (see Publius, pp. 214–15, paper 35).[24]

According to Publius, a coherence of interests not only exists on a national scale in America. That coherence of interests also defines a quantitatively greater public good in union than in disunion. This is the argument of paper eleven, where he demonstrates how a united commerce will be much more prosperous than a collection of disunited commerces (see Publius, pp. 89–90, paper 11).

These arguments are precisely the ones a republican statesman would make in trying to adapt the republican tradition to a larger, and increasingly commercial, society. Unfortunately, these arguments are not immediately persuasive. The fact that commerce raises the value of land is not an unmixed blessing for farmers, nor does it necessarily mean an increase in the relative prices of their products. Similarly, the fact that commerce increases national prosperity does not guarantee farmers a share in that increase; the claim that American agriculture and commerce are each internally homogeneous assumes they are mutually heterogeneous; and the assertion of a harmony between those two mega-interests with respect to relying on tariffs as a source of government revenues is much more reassuring to one of them. We, finally, should recall that the argument of paper ten characterizes both interests as *internally* heterogeneous by pointing out the obvious distributional effects of many tax policies (and, we might add, treaty provisions, even though this seems to be an easy case for Publius) (see Publius, pp. 79–80, paper 10).

In part, Publius's difficulties here are intrinsic to the traditional republican idea of the public good. By extending the republican sphere, however, he undoubtedly has further stretched the credibility of that idea. As a relative matter, it is more plausible on the state level because its underlying premise—social homogeneity—is more plausible. Publius can only recognize this fact and, in order to uphold his nationalist goals, seek to make a virtue out of a necessity. He must find merit in diversity, as he does in fashioning an antidote to factions in paper ten and as he does in celebrating the more prosperous commerce the union promotes in paper eleven. These pluralistic moves appear to have been imposed on Publius by his uncompromising commitment to the union.

Even so, we should not discount the extent to which Publius did struggle to justify that commitment on traditional republican grounds. He was also committed to an identifiably republican tradition of ideas. His prediction that interstate differences will decrease over time should be read in light of an attempt to balance his commitments to nationalism and traditionalism (as well as social realism). This prediction, twice repeated, suggests that he still shared, at least to some extent, the traditional republican concern for social homogeneity (see Publius, p. 334, paper 53; pp. 348–49, paper 56; pp. 367–68, paper 60). A second, more convincing way he defined the union as a cohesive national republic supports the interpretation that he revised and rearranged traditional republican beliefs into a new theoretical synthesis rather than simply discarded them.

A Union of Sentiments

Above all else, Publius defines America as a patriotic community. He stresses its ideological homogeneity. A certain configuration of public opinion makes the union the most natural political society for the American people. Their own fervor for the union constitutes an additional way it is a republican society, and a better one than the individual states.

Publius's emphasis on patriotism represents both a reformulation and a relaxation of the traditional republican concern for social homogeneity. It replaces relatively tangible, economic measures of homogeneity with more amorphous, noneconomic ones. It represents a similar shift on civic virtue. This shift is from a more to a less participatory politics; from more to less overt modes of citizen support for the regime. Again, these pluralistic modifications of traditional republican beliefs seem incumbent on Publius once he has extended the sphere of republican government.[25]

While Publius, in paper ten, contends that size can help secure a virtuous political leadership, he never suggests that it can help secure a virtuous citizen body (see Publius, pp. 83–84, paper 10).[26] He seems both to partially accept the traditional republican argument that civic virtue can only be generally effectual in small political societies and to partially transform that argument by claiming that patriotism, something akin to civic virtue in its psychological dimensions, can be generally effectual in large societies. Publius, though, does not just substitute patriotism for civic virtue. He, on the one hand, still appears to be very concerned with the institutionalization of civic virtue in a political elite. On the other hand, he appears to be somewhat dubious about the capacities of most citizens to attain that level of political sophistication and to be firmly convinced that more passive or deferential modes of citizen behavior are ultimately advantageous anyway.[27] Presumably, then, he considered an emphasis on patriotism not so much the cost of national republicanism as one of its assets.

Paper two most strongly evokes Publius's stress on ideological homogeneity. Employing compact yet emotionally charged language, he defines the union as America's most natural society in that paper. His definition has four aspects. First, the American territories are naturally connected, as if designed by Providence to be one nation. Second, the residents of those territories are naturally connected. Americans form "a people descended from the same ancestors, speaking the same language, professing the same religion, attached to the same principles of government, very similar in their manners and customs." Third, Americans are still more tightly joined by their common experiences, especially those incurred during the war of independence. The fourth tie that binds is how most Americans share Publius's patriotic sentiments. He underscores the importance of this last factor by opening and closing the paper with a spirited defense of his audience's patriotism (see Publius, pp. 37–38, 41, paper 2).

Publius often affirms his confidence in the fundamental patriotism of the American people, distinguishing it as the most powerful bond of union (see Publius, p. 36, paper 1; p. 39, paper 2; p. 42, paper 3; pp. 103–4, paper 14; pp. 517–18, paper 84).[28] This motif does not merely constitute an effort to place that sentiment squarely on the side of the Constitution or to cultivate the underlying strength of the union against what he sees as the incipient forces of disunion. It also is an attempt to define the union as a national republic. Patriotism gives the union a cohesiveness in sentiments which it may well lack in interests. Because that sentiment is a prevailing one, the union forms not only America's most natural society but its best republican society.

Judged by traditional republican standards, however, patriotism provides a relatively weak basis of citizenship. This attenuation is clearly revealed in paper forty-nine, where Publius rejects Thomas Jefferson's proposal that "separation of powers" disputes be settled by popular referenda, even though that proposal "seems strictly consonant" with "republican theory" (Publius, pp. 313–14, paper 49). It is also clearly revealed in his recurrent admonitions against an "excessive" republican jealousy (see Publius, p. 170, paper 26; p. 196, paper 32; pp. 298–99, paper 46; p. 345, paper 55; p. 395, paper 64).[29]

Nonetheless, a now-familiar framework for analyzing Publius's political thought remains compelling. This framework focuses on the in-betweenness or transitional character of his thought. In this context, it advises us to exaggerate neither the participatory nature of traditional republican views of citizenship nor the nonparticipatory nature of Publius's views. He seems to be steering a course between the participatory thrust of traditional ideas of civic virtue and republican jealousy and the more deferential citizen ethic suggested by the traditional idea of a natural aristocracy. It is true that he veered farther in a nonparticipatory direction than did many other American statesmen at the time. Whatever Jefferson's (or the Federal Farmer's) republican theory prescribed, Publius's did not call for periodic appeals to the people. Yet, in paper forty-nine he does not completely reject popular referenda, as in ratifying and amending constitutions (see Publius, p. 314, paper 49). Nor, in warning of the dangers of an excessive republican jealousy, does he fail to appreciate the value of the well-adjusted republican jealousy he detects in most American citizens (see Publius, p. 260, paper 41; p. 353, paper 57; p. 395, paper 64). He would have differences on these two dimensions understood as differences in degree and as a matter of discovering the proper balance between them. A less participatory republicanism, though, was also inherent in a credible national republicanism. Here, Publius, again, made a virtue out of a necessity. He summoned the traditional idea of a natural aristocracy to prove that, within limits, a less participatory government is a better one. Because of (rather than in spite of) the fact that the federal government will be less participatory than the state governments, he can comfortably, on traditional republican grounds, recommend a transfer of powers to the federal level.[30]

Publius constructs a massive scaffolding in the fourth head to defend the less frequent elections, more indirect elections, and smaller representative bodies which will exist on the federal, as compared to state, level under the Constitution.[31] This scaffolding rests on the supposition that less participatory institutions are necessary to maintain a natural aristocracy in government. At the same time, Publius argues that these institutions are sufficiently participatory so as not to violate the spirit of republican government. He insists that any differences with his opponents on political structure lie safely inside the bounds of republican theory.

In the middle of a series of papers on the House of Representatives, Publius pauses to make another one of those grandiloquent pronouncements that are scattered throughout his work.

> The aim of every political constitution is, or ought to be, first to obtain for rulers men who possess most wisdom to discern, and most virtue to pursue, the common good of the society; and in the next place, to take the most effectual precautions for keeping them virtuous whilst they continue to hold their public trust. (Publius, p. 350, paper 57)

Although the Anti-Federalists would not object to the explicit claims of this statement, they would be disarmed by its tenor. Is not the intention behind obtaining such rulers to preclude most American citizens from having any influence on the government? Publius, however, goes on to emphasize that the primary precaution for "keeping them virtuous" is their electoral "responsibility to the people." He, thus, emphasizes the popular means to a goal which only seems "oligarchic" to Americans whose "flaming zeal for republican government" will not allow them to trust their fellow citizens to choose good rulers (Publius, pp. 350–51, paper 57).[32]

The most revealing case, though, remains his rejection of Jefferson's proposal for popular referenda on "separation of powers" disputes. This case vividly contrasts two different conceptions of citizenship.[33] Publius's principal objection to the proposal is that it would undermine popular support for the federal government. Institutionalizing such a process of periodic appeals to the people would unsettle the moorings of any government. It is a case of too much participation, which no viable republican theory would prescribe (see Publius, pp. 314–15, paper 49).[34]

What is at stake for Publius in this case, however, is not simply popular support for the federal government. What is also at stake is popular support for the federal as opposed to state governments and, at a deeper level, patriotic as opposed to localist sentiments. If the union is primarily defined as a patriotic community, then the critical problematic for national republicans is divided loyalties, not clashing interests.[35]

Throughout his consideration of this problematic, Publius assumes that Americans are a patriotic people and with time will become more so. The cogency of this assumption, though, depends on the ratification of the Constitution. Publius predicts that "in proportion as the United States assume a national form and a national character, so will the good of the whole be more and more an object of attention" (Publius, p. 395, paper 64).[36] He believes the division of government powers will have a significant impact on the divided loyalties of the American people because only an energetic federal government will be able to make itself venerable. Yet, he also insists that the federal balance of power will be decided as much by the American people as by the Constitution (see Publius, p. 197, paper 31; p. 294, paper 46). He postulates an open-ended, dynamic relationship between structural and psychological factors. The Constitution, in strengthening the federal government, will intensify the patriotic community, which, in turn, will further strengthen the federal government.

Publius, nevertheless, gives greater credence, at least rhetorically, to the opposite scenario of a federal balance that favors the state governments, creates veneration for those governments, and nurtures localist sentiments. He presents this scenario in paper seventeen, where he argues that the state governments will enjoy more popular support than the federal government because they are closer to the people. This advantage is based on a general principle of human nature.

> It is a known fact in human nature that its affections are commonly weak in proportion to the distance or diffusiveness of the object. Upon the same principle that a man is more attached to his family than to his neighborhood, to his neighborhood than to the community at large, the people of each State would be apt to feel a stronger bias towards their local governments than towards the government of the Union . . . (Publius, p. 119, paper 17)

The comparative advantage of the state governments in popular support will be augmented by their responsibility for the administration of justice. This second advantage follows from a related, but still distinct, principle of human nature.

> . . . the ordinary administration of criminal and civil justice . . . being the immediate and visible guardian of life and property, having its benefits and its terrors in constant activity before the public eye, regulating all those personal interests and familiar concerns to which the sensibility of individuals is more immediately awake, contributes more than any other circumstance to impressing upon the minds of the people, affection, esteem, and reverence towards the government. (Publius, p. 120, paper 17)[37]

However, Publius implicitly removes the linchpin of this argument in the next group of papers. In the third head, he indicates how the federal government will possess the constitutional means to make itself more venerable than the state governments. The federal government's exclusive responsibility for the national defense will be extremely important in this transformation (see Publius, pp. 208–11, paper 34). But Publius most directly undercuts the argument of paper seventeen in paper twenty-seven. In that paper, he suggests the federal government will also benefit from the second, if not the first, principle of human nature enunciated in the earlier paper. He, furthermore, expands the criteria of popular support. An additional criterion issues from yet a third general principle of human nature, "that their [citizens'] confidence in and obedience to a government will commonly be proportional to the goodness or badness of its administration." Since Publius fully expects the federal government to be better administered than the state governments, he must also expect it to be more venerated (see Publius, pp. 174–76, paper 27).[38]

Publius, consequently, advocates a shift in the criteria of good government away from its closeness to the people toward its performance. He also suggests the two criteria are, to a certain extent, antithetical. Governments can be too close to the people, something his opponents seem unable to comprehend. To return once more to his rejection of

Jefferson's proposal for popular referenda on "separation of powers" disputes, Publius questions how the federal government can be thought to be well administered under such a scheme; indeed, he questions how it *can* be well administered under such a scheme (see Publius, p. 314, paper 49). Later, he discloses how Jefferson's proposal is based on a counterfactual, that the internal structure of the federal government, in particular its separation of powers, is defective. Publius insists that the separation of powers is, to the contrary, craftily designed to boost government performance (see Publius, pp. 322–23, paper 51).

This criteria-shift is an integral part of Publius's case for a change in the scale of republican government. Only according to the new criterion can Americans judge the federal government superior to the state governments. Publius, however, argues that Americans *will* judge the federal government superior because the criterion of effective administration is immanent in mass psychology. The prevalence of that attitude is essential to the existence of the union as a national republic, and to its choiceworthiness over a federal republic or several disunited republics.

Publius, though, did not advocate a complete shift away from the traditional criterion of closeness to the people insofar as he claimed that large-scale governments could develop a sufficient psychological proximity to distant citizens through an effective administration. Besides, he could argue that the latter criterion, which he felt clearly favored large-scale governments, was equally republican. He pointed out the great danger of confusing means and ends in thinking about the criteria of good government, daring the Anti-Federalists to confess that they preferred the state governments to the federal government just because the state governments were closer to the people (see Publius, p. 289, paper 45).[39]

No matter what arguments he might have made, Publius did, and could, not establish the superiority of large-scale governments and societies solely on traditional republican grounds. Nonetheless, his efforts were very successful on their own terms and, to that extent, he turned conventional wisdom on its head. Conventional wisdom says that "almost every one of these states" (except Rhode Island?) is too large to be a republic (Publius, p. 73, paper 9). Publius's new wisdom says that all the states, especially Rhode Island, are too small. After questioning Rhode Island's potential as an independent republic, he makes the startling claim that "the larger the society, provided it lie within a practicable sphere, the more duly capable it will be of self-government" (Publius, p. 325, paper 51).

That "practicable sphere" is defined, actually redefined, in paper fourteen, a paper which as much as ten sets the tone for Publius's boldness.

Paper Fourteen

This paper, the last one in the first head, opens with a summary of the argument of that head.

> We have seen the necessity of the Union as our bulwark against foreign danger, as the conservator of peace among ourselves, as the guardian of our commerce and other common interests, as the only substitute for those military establishments, which have subverted the liberties of the old world, and as the proper antidote for the diseases of faction, which have proved fatal to other popular governments, and of which alarming symptoms have been betrayed by our own. (Publius, p. 99, paper 14)

Once he has vindicated the large-scale union by these widely shared ends, Publius explicitly considers the Anti-Federalists' size objection. He first denounces it as "a prevailing prejudice" and contends that it rests on the error of confounding republics and democracies. He repeats his almost eager concession (of papers nine and ten) that the turbulent experiences of the small Greek republics are indefensible against those who would use them to besmirch the republican cause. However, he now notes that those city-states were really democracies, not republics. The ancient Greeks were not cognizant of "the great principle of representation." We are. Americans, in fact, have the distinction of bringing that principle to perfection. As Publius has also already demonstrated (in paper ten), the representative principle not only permits extended republics but its perfection consists in extended republics (see Publius, pp. 100–101, paper 14).[40]

Publius next presents the "true" limits on the size of republics. Those limits are defined by "that distance from the center which will barely allow the representatives of the people to meet as often as may be necessary for the administration of public affairs." This definition could only have provoked howls of disbelief from his opponents. Yet, it does serve to enfranchise the union as a republic. The union, after all, is not much larger than several European nations which successfully practice the principle of representation. In this context, he does not mention his personal belief that none of those countries is a republic, though he does mention, only to dismiss as a premature consideration, the fact that the United States' unorganized territories make it much larger than each of those countries. He then observes that federalism also permits extended republics, again using what is in dispute to his rhetorical advantage (see Publius, pp. 101–2, paper 14).[41]

The paper closes on a different note. Publius now acknowledges the validity of his adversaries' claim that a (nonfederal?) republic as extensive as the union is unprecedented. He just denies its practical force.

> But why is the experiment of an extended republic to be rejected merely because it may comprise what is new? Is it not the glory of the people of America that, whilst they have paid a decent regard to the opinions of former times and other nations, they have not suffered a blind veneration for antiquity, for custom, or for names, to overrule the suggestions of their own good sense, the knowledge of their own situation, and the lessons of their own experience? (Publius, p. 104, paper 14)

"The leaders of the Revolution" followed this method. In erecting republican governments in the states, "they pursued a new and more noble course . . . they reared the fabrics of governments which have no model on the face of the globe" (Publius, p. 104, paper 14). Publius challenges their successors—his audience—to follow the same method in perfecting the federal system.[42]

If Publius's own political methodology seems to be characterized more by experimentation than by a "decent regard" for tradition, this impression must be tempered by other passages in which he is more solicitous of his precedents and in which he recommends a more cautious, experiential approach to constitutional problems.[43] Still, it remains true that Publius, when compared to such Anti-Federalists as the Federal Farmer, was conspicuous for his theoretical audacity. He was much more likely to appeal to the uniqueness of American experience than to "the opinion of former times and other nations." His stance throughout paper fourteen is a defiant one. He, in effect, accuses his opponents of intellectual cowardice. They are being disloyal to the "manly spirit" and missionary impulses of the American people, to which "posterity will be indebted for the possession, and the world for the example of the numerous innovations displayed on the American theater in favor of private and public happiness" (Publius, p. 104, paper 14).[44]

Publius's boldness must, in particular, be measured against the republican tradition. By consistently portraying the union as America's best possible republic, he can justifiably argue that his own innovations are innovations for the sake of that tradition, not in violation of it. He can also insist that under American conditions the rashest experiment in republicanism is disunion, not union. He, finally, can proselytize against disunion in the hallowed tones of republican declension, warning that "the kindred blood which flows in the veins of American citizens, the mingled blood which they have shed in defense of their sacred rights, consecrate their Union and excite horror at the idea of becoming aliens, rivals, enemies" (Publius, p. 104, paper 14).

We strongly suspect that disunion was more horrible to Publius on nationalist than on republican grounds. For him, though, those two grounds were not distinct. To a remarkable degree, he was able to integrate his commitment to the union into America's prevailing tradition of ideas. Yet, the resulting synthesis does appear as an adulteration—from our perspective, a pluralistic adulteration—of republicanism by nationalism.

More generally, the Federalists and "their" Constitution unleashed (as they had embodied) nationalistic forces that quickly overwhelmed the opposition.[45] These same forces also stretched the republican tradition thinner and thinner. The states-rights position would never again be as "pure" as it had been in Anti-Federalism. This dilution, however, did not mean it would never again be as powerful. The states-rights position remained vigorous precisely because of its impurities, which allowed it to keep pace with changing standards of political relevance. But it also remained vigorous because of its persisting republican assumptions, which continued to appeal to many Americans.

PART TWO

THE NULLIFICATION DEBATE

To a significant extent, the nullification debate was a replay of the ratification debate. Webster reenacted Publius's national-republican role and Calhoun, the Federal Farmer's federal-republican role. Here was remarkable continuity, expressed both as a single tradition of political ideas and through the basic division within that tradition. Yet, beneath the surface, much had changed. The Constitution had been ratified and opposition to it had, almost overnight, become a political heresy. At a deeper level, the nature of America's prevailing intellectual tradition, social realism, and nationalism had all continued to evolve, dialectically, in a pluralist direction. Americans had also developed a strong legal tradition that promised to anoint the victor of this new debate. None of these dimensions of the debate, however, was self-interpreting. If the drama of the first debate lay in the background of a popular referendum on a new constitution, Calhoun and Webster themselves provided the spectacle of the second debate as they fought for the soul of that Constitution on the floor of the United States Senate.

5

John C. Calhoun, South Carolina, and the Union

John C. Calhoun bears the patrimony of the Anti-Federalists. Except, he argues as if they had written the Constitution. His reading of the ratification debate can only be considered odd. He did not find much support at the time for either his historical or constitutional interpretations.[1] Yet, his understanding of the Constitution is similar to the Federal Farmer's "alternative Constitution," and his backward glance at American political development mirrors the course the Federal Farmer feared it would take. They differ in that the Federal Farmer warned that consolidationist declension would naturally flow from the Constitution; Calhoun portrays it as an irruption and sounds the restoration theme.[2]

Calhoun implicitly divides the 1787–88 disputants into three parties: Anti-Federalists, Federalists, and consolidationists. He claims that the second group emerged victorious from the Philadelphia convention, the key decision being the one *not* to grant Congress a veto over state laws. He translates that decision into a tacit recognition of the states' right to veto federal laws. According to Calhoun, the founding fathers would have endorsed South Carolina's nullification ordinances. His historiography makes the Federalists more "federal" than the real Anti-Federalists; nullification is more extreme than the state checks the Federal Farmer wanted to insert into the Constitution (see Calhoun II:300–301).[3]

Paradoxically, the Federal Farmer, who was much more critical of the Constitution, was closer to it in spirit. In 1833, though, it would have been politically suicidal for Calhoun to criticize the Constitution and to present himself as an intellectual heir of the Anti-Federalists. Nevertheless, the nullification crisis was a clear case where the undeniable power of the Constitution to structure political debate in America did not preclude serious conflict between different interpretations of the Constitution.[4]

Notwithstanding their different notions of federalism and their different polemical situations, the area of agreement between the Federal Farmer and Calhoun remained large. They both looked at American constitutional problems from an identifiably republican perspective and they both came to the same basic conclusion. American conditions mandate a strongly federal system of government because the individual states better fit traditional republican beliefs than the union does. These were the primary presumptions and purposes behind Calhoun's nullification doctrine, just as they were behind the Federal Farmer's federal balance.

Calhoun's doctrine, however, betrays several countervailing tendencies. First, Americans (including himself) are firmly committed to the union; rather than sowing the seeds of disunion and anarchy, nullification will actually cement the union. Second, the union is (or can be) a national republic, at least in an attenuated sense; the federal government is not, and should not be, a strictly federal one.[5] Third, it is doubtful whether traditional republican beliefs are even realistic on the state level; Americans' major political assumptions cannot be strictly republican ones. If not the first two, nationalistic tendencies, then the third, pluralistic tendency is more pronounced in Calhoun than in the Federal Farmer. Still, these tendencies were countervailing to a more fundamental federal republicanism for both the Federal Farmer and Calhoun.[6]

Other ways of accounting for Calhoun's position appear less significant. To an even greater extent than the ratification debate, the nullification debate raised the democratic issue in a form which subsumed it under the federal issue. Calhoun's concern was the possible conflicts between national and state or sectional majorities. Thus the problem of majority tyranny was to be addressed through the federal structure, not the internal composition of the federal government.[7]

Scholars have more often turned to the slavery issue in an attempt to explain Calhoun's position. Nullification doctrine would offer ironclad protection for the South's peculiar institution from any federal efforts to abolish slavery. Calhoun identifies the recent emancipation debates in the Virginia legislature as one of the forbidden fruits of the opposing doctrine of political consolidation. He also rebukes Senator John Forsyth (D-Georgia) for suggesting that the republican guarantee clause of the Constitution could be invoked against the government of South Carolina in the present controversy for its alleged mistreatment of the state's unionist minority. He is dumbfounded that a senator from Georgia could even mention the clause when it opens the door for Northern politicians to invoke the clause against the existence of slavery in the Southern states. At one point in his first nullification speech, he more or less says that slavery is at stake in the nullification crisis, for it is a crisis "in which the weaker section, with its peculiar labor, productions, and institutions, has at stake all that can be dear to freemen" (Calhoun II:261).[8]

In the course of reprimanding Forsyth, Calhoun, nonetheless, professes to believe that "there are now no hostile feelings combined with political considerations, in any section, connected with this delicate subject" (Calhoun II:309). The only senator to say otherwise—Gabriel Moore (D-Alabama)—was totally ignored by the other senators who spoke during the debate. Most of them, including Calhoun and Webster, never explicitly refer to slavery.[9] Although that "failure" is hardly conclusive, the claim that slavery was the central issue of the nullification crisis—before the congressional debates over the gag rule and abolitionist mailings, the annexation of Texas, the Mexican War, and the Wilmot Proviso—seriously foreshortens the march of events. We need only compare Calhoun's nullification speeches with his increasingly strident, proslavery speeches of the late 1830s and 1840s to realize that slavery was still the repressed issue it had been during the ratification debate, even if it now was closer to the surface.[10]

Another narrative of the nullification crisis stresses its broader economic aspects. The crisis pitted Southern agriculture against Northern commerce. Calhoun did assume the role of sectional spokesman during the debate, trying (for the most part, unsuccessfully) to rally other Southern senators to South Carolina's cause (see Calhoun II:223, 261–62, 308–9). He insisted that nullifying protective tariffs was the only way of halting the northward flow of capital which resulted from such tariffs because simple majorities could always be attained on tariff bills through logrolling. If he argued that protective tariffs violated the Constitution as well as the South's economic integrity, we naturally wonder whether the first argument was merely intellectual scaffolding for the second (see Calhoun II:198, 219, 221, 223–24, 236, 240).[11]

Yet, Calhoun himself downplays the economic issues for the "higher" constitutional issues involved in the controversy. Or, rather, he claims that is what the Jackson administration has done by introducing the Force Bill. That bill "has merged the tariff, and all other questions connected with it, in the higher and direct issue which it presents between the federal and national system of governments" (Calhoun II:307). Moreover, Calhoun, the president, and almost all the other senators—with Webster as the notable exception—were willing to compromise the economic issues but not the constitutional ones. This, I think, bespeaks the rough priority between the two sets of issues in their own minds. The crisis, of course, was "solved" by the majority in just that manner, with passage of the "consolidationist" Force Bill and *then* the Compromise Tariff of 1833.[12]

For Calhoun, if not for the others, a debate over the nature of the American political system points to still higher issues. Near the middle of his first speech, he pauses to summarize:

> In reviewing the ground over which I have passed, it will be apparent that the question in controversy involves that most deeply important of all political questions, whether ours is a federal or a consolidated government;—a question, on which the decision of which depend, as I solemnly believe, the liberty of the people, their happiness, and the place which we are destined to hold in the moral and intellectual scale of nations. (Calhoun II:238)

Calhoun, perhaps uniquely in American history, treats the question of the best constitutional means to the liberal ends as a "pure theory" question. In a passage that is all too easy to dismiss as personal braggadocio, he takes umbrage at Senator John Clayton's (NR-Delaware) charge that his doctrine is "metaphysical"; not to deny the charge but only its pejorative implications.

> If by metaphysics he means that scholastic refinement which makes distinctions without difference, no one can hold it in more utter contempt than I do; but if, on the contrary, he means the power of analysis and combination—that power which reduces the most complex idea into its elements, which traces causes to their first principle, and, by the power of generalization and combination, unites the whole in one harmonious system—then, so far from deserving contempt, it is the highest attribute of the human mind . . . And shall this high power of the mind, which has effected such wonders when directed to the laws which control the material world, be for ever prohibited, under a senseless cry of metaphysics, from being applied to the high purpose of political science and legislation? I hold them to be subject to laws as fixed as matter itself, and to be as fit a subject for the application of the highest intellectual power . . . the time will come . . . when politics and legislation will be considered as much a science as astronomy and chemistry. (Calhoun II:232–33)[13]

Calhoun's "scientific" solution to the fundamental problem of politics and legislation is the theory of the concurrent majority. On any issue which "naturally" divides a political society into two, the final decision should require the consent of a majority of the units in both the majority and the minority. (These units may be individuals, interest groups, or member-states in a federation; acting for themselves or through deputies in a representative assembly.) This particular method of constructing majorities maximizes the liberty and happiness of each of the units, and hence of the whole. Calhoun disclaims any originality for his theory, contending that it is embodied in the amending power of the Constitution. As an appeal to that power, the nullification of a federal law by one or more states on the grounds of unconstitutionality triggers the concurrent majority of states underlying the American system of government. If the nullifying state(s) loses its appeal, then it can, as a last resort, secede from the union. These precepts follow from the fact that the Constitution is a compact between once and still sovereign states, ratified by a concurrent majority of them. Calhoun's nullification doctrine, therefore, sets up a deductive argument from pure theory to applied theory to a political practice, from the concurrent-majority principle to American constitutional principles to the nullification of federal laws by sovereign states (see Calhoun II:250, 255, 260, 301–2).[14]

However, we are likely to misconstrue Calhoun's doctrine if we fail to recognize that the principal assumptions of his applied theory are federal-republican ones. The American states are not themselves governed by concurrent majorities because they are republics. The union is, because it is not itself a republic.[15] Calhoun's applied theory becomes a question of scale. Aside from its metaphysical quirkiness, his nullification doctrine resembles the traditional small-republic argument. While that resemblance is not the last word on his doctrine, it is the first and most important word.[16]

In the end, we must be impressed with how much Calhoun's political rhetoric echoes the Federal Farmer's, and less because of the constitutional arguments than the underlying republican assumptions. We must especially be impressed because of the dramatic social changes which have transformed the American landscape during the intervening years. These changes may well explain Calhoun's greater nationalistic and pluralistic tendencies. But only the persistence of an identifiably republican tradition of ideas can account for the basic continuities in their rhetoric.

The Republic of South Carolina

At least for purposes of the nullification debate, South Carolina is Calhoun's concrete republic. It was natural that he should defend the actions and honor of his state against the incriminations of other senators, particularly when he was so intimately involved in those actions. His apology, though, is revealing for its recurrent republican language, both, positively, on the state level and, negatively, on the national level.

Significantly, the first accusation Calhoun addresses is not a constitutional one. It is the accusation that South Carolina's nullification ordinances are motivated by avarice, that in nullifying the tariffs her citizens are simply seeking to avoid paying their fair share of the public burdens.[17] Calhoun's immediate response is to glorify his state's contributions in blood and money to the national cause ever since the outbreak of the war of independence. Later, he insists that South Carolina supported the Tariff of 1816 out of "disinterested motives" as her "obvious policy" was (and is) free trade. But now, after a string of progressively higher protective tariffs, he claims that it is no longer just a question of economic policy. "The great principles of constitutional liberty" are at stake and South Carolina's "brave sons" are willing to make the ultimate sacrifice to uphold those principles. No, it is the rest of the country, not South Carolina, which "has sunk into avarice and political corruption" (see Calhoun II:198, 212–13, 229, 235).[18]

Calhoun, secondly, confronts the charge that South Carolina's stance is based on "passion and delusion."[19] He does so in a remarkable series of paragraphs in his first nullification speech which describes how his state came to take the fateful step of nullifying the tariffs.

> Never was there a political discussion carried on with greater activity, and which appealed more directly to the intelligence of a community. Throughout the whole, no address has been made to the low and vulgar passions . . . In this great canvass, men of the most commanding talents and acquirements have engaged with the greatest ardor; the people have been addressed through every channel . . . No community, from the legislator to the ploughman, were ever better instructed in their rights; and the resistance on which the state has resolved, is the result of mature reflection accompanied with a deep conviction that their rights have been violated, and that the means of redress which they have adopted are consistent with the principles of the constitution. (Calhoun II:214–15)

Far otherwise does Calhoun characterize the relationship between leaders and followers on the national level. From Washington, the public is "instructed" by "a great, a powerful, and mercenary corps of office-holders, office-seekers, and expectants, destitute of principle and patriotism, and who have no standard of morals or politics but the will of the Executive—the will of him who has the distribution of the loaves and the fishes" (Calhoun II:253).

Although the party game is being actively, and shamelessly, played in the federal government, South Carolina has attempted to isolate herself from that festering source of corruption. Calhoun even traces the vehemence of the attacks on South Carolina's motives to her refusal to participate in the recent presidential election.

> And why has she thus been assailed? Merely because she abstained from taking part in the Presidential canvass—believing that it had degenerated into a mere system of imposition on the people—controlled, almost exclusively, by those whose object it is to obtain the patronage of the Government, and that without regard to principle or policy. (Calhoun II:207)[20]

South Carolina, consequently, stands vindicated on republican grounds. Her citizens are intelligent and virtuous, her public officials form a conscientious natural aristocracy, and her politics is unblemished by party divisions. Conversely, the union (excluding South Carolina, of course!) stands condemned on the same grounds. Her citizens have been corrupted by rapacious national parties, her public officials would fail any test of statesmanship, and her politics is fractured by quarrels over "the spoils of victory." There, however, is one troubling anomaly in this extraordinary contrast—the South Carolina unionists.

Specifically, Calhoun must answer the allegation that the test oath, which requires South Carolinians to swear to support their state's nullification ordinances, oppresses those who opposed that line of policy.[21] He contends that in practice the unionists are not being oppressed because the oath is only being administered to state officials who might be involved in implementing the ordinances. The more important point, though, is that the unionists agree with the nullifying majority that protective tariffs are "unconstitutional and oppressive." The division between South Carolina unionists and nullifiers does not belie an intrastate unity of interests and opinions. Nor, is it a party split. It, rather, is merely a minor difference of opinion over the best remedy for a common grievance. Calhoun presses the point home by noting that outside critics overlook the fact that the oppression South Carolina, including her unionist minority, faces within the union from the federal tariffs and Force Bill dwarfs the oppression that the minority faces within South Carolina from the nullification ordinances and test oath (see Calhoun II:222–26).

Calhoun suggests that those critics also mistake the nature of the American political system. The fact of the matter is that individual citizens or minorities of them do not enjoy the same rights within states as individual states or minorities of them enjoy within the union. Because states are sovereign communities, individual citizens are bound to obey state majorities but individual states are not bound to obey national majorities, at least not simple ones. In other words, the states are governed by absolute majorities of citizens and the union, by concurrent majorities of states (see Calhoun II:221–22, 260).

This difference in majority rules has the force of an "ought" for Calhoun. He, after all, does not think that the federal system is currently operating on concurrent-majority principles. On one level, the force of this "ought" lies in the claim that the system was originally intended to operate on such principles (see Calhoun II:252–53). Its force, however, goes deeper than constitutional principles. The difference in majority rules is also dictated by the difference in the aggregate qualities of the individual states and the whole union. Simply stated, the states fit traditional republican assumptions; the union does not. Because, for the most part, those are Calhoun's political assumptions and that is his social analysis, the application of concurrent-majority theory to American conditions results in a federal-republican theory.

Calhoun's fundamental federal republicanism explains how he can characterize the nullification controversy as a contest between liberty and union (see Calhoun II:264).[22] A conflict between those two treasured American values arises in his mind because he believes liberty is *in* the states, not the union.

It is clear that the liberty which Calhoun places at the heart of the crisis is the freewheeling pursuit of economic interests. In the course of denying the "avarice" charge, he, thus, declares that: "The real question at issue is: Has this Government a right to impose burdens on the capital and industry of one portion of the country, not with a view to revenue, but to benefit another?" He goes on to show just how much a succession of protective tariffs has adversely affected the interests of his fellow citizens, concluding that it has imposed "a ruinous burden on the labor and capital of the State, by which her resources are exhausted—the enjoyments of her citizens curtailed—the means of education contracted—and all her interests essentially and injuriously affected" (Calhoun II:198–99).

Throughout Calhoun's first nullification speech, the very existence of an union of diverse economic units appears as a threat to the well-being of each unit. He seriously calls into question the possibility of a national harmony of economic interests. He holds out little hope that *any* federal policy can be fashioned which does not heavily handicap at least one of the units. In contrast, he never seems to doubt that such policies, such a public good, because such an integrated economy, exist in the states. They are economic units; the liberty and happiness of the American people are secure in their state communities. State homogeneity is Calhoun's controlling assumption when he applies concurrent-majority theory to the American case, just as it was the controlling assumption of prior federal-republican theories (see Calhoun II:250).

Nonetheless, the size of the threat the union presents to the aggregate welfare of the American people depends on the structure of the union. Calhoun can, alternatively, portray himself as defending liberty and union; that is, the true, federal union which respects its constituent-parts as the loci of American liberty as against the false, consolidated union which does not (see Calhoun II:301).[23]

Both as a legal and sociological fact, Calhoun claims that the true union is a federal one. The "big lie" of his opponents, the putative defenders of the union, is that it is "one great community" and the states, "mere fractions or counties" (Calhoun II:228).[24] For a number of years, the American system of government has falsely operated on this fiction as an absolute-majority system. The consequences have been disastrous for the stability of the union, not to mention the interests of most American citizens. According to Calhoun, nullification is the only way to arrest this decline and to force the system back into its original concurrent-majority mode. He thinks that it is the only way of saving a union worth saving (see Calhoun II:256).

In these terms, Calhoun can also champion his own patriotism. He is prepared to vindicate his "consistency of conduct, purity of motives, and devoted attachment to the country and its institutions" (Calhoun II:265).[25] Given his theoretical assumptions, though, this attachment must be an ambivalent one; not because he would ever counsel disunion but because his loyalties (as well as those of most other Americans) must be powerfully divided between the union and its institutions and the states and their institutions. His reply to Clayton's appeal to their common citizenship exposes this ambivalence.

> The Senator from Delaware, as well as others, has relied with great emphasis on the fact that we are citizens of the United States. I do not object to the expression, nor shall I detract from the proud and elevated feelings with which it is associated; but I trust that I may be permitted to raise the inquiry, In what manner are we citizens of the United States?, without weakening the patriotic feeling with which, I trust, it will ever be uttered. If by citizens of the United States he means a citizen at large, one whose citizenship extends to the entire geographical limits of the country, without having a local citizenship in some State or territory, a sort of citizen of the world, all I have to say is, that such a citizen would be a perfect nondescript. (Calhoun II:242)[26]

Calhoun's defense of the federal union provides a paradoxical basis for his nationalism. It submerges national loyalties in state loyalties.[27] Yet, any stronger nationalist claims Calhoun might make would collide with his own fundamental theoretical assumptions and empirical observations. These are the assumptions and observations of a federal republican. He, nevertheless, does not completely dismiss the prospects of an American national republic, any more than the Federal Farmer did. This aspect of nullification doctrine discloses the hold of American nationalism even on the "arch-nullifier."[28]

A National Republic, If . . .

Calhoun offers the concurrent-majority principle as an elixir which not only will restore the federal system to its original purity and prevent disunion but also will transform the union into a national republic. It will invest the union with many of the republican qualities which seem to naturally inhere in smaller political societies; in particular, in South Carolina. In this sense, the union currently is not a national republic only because of the abuses of the absolute-majority principle.

Again, this aspect of Calhoun's doctrine is revealed through his responses to a series of objections to South Carolina's nullification ordinances, this time with respect to their probable consequences.

The worst-case scenario—yes, even for Calhoun—is that nullification will lead to disunion. He, however, contends that it, in fact, is "the strongest cement" of the union (Calhoun II:257).[29] As compared to absolute-majority rule, concurrent-majority rule creates a greater attachment to the whole because it gives each part a more decisive voice in the policies of the whole. Calhoun's counterargument relies on a reading of Roman history according to which its concurrent-majority principle—the dual assent of patricians (senators) and plebeians (tribunes) to major policy decisions—did not foment disunion. To the contrary, it produced the most powerful, because least divided, republic of ancient times. At the end of this history lesson, Calhoun proposes a universal law to the effect that "the virtue, patriotism, and strength" of a state are directly proportionate to "the perfection of the means of securing such [a concurrent] assent" (Calhoun II:259).[30]

If the Roman example proves that the concurrent-majority principle can diffuse virtue and patriotism among the populace, it also proves that the principle can distill a natural aristocracy out of the populace. Calhoun postulates such a tendency in rebutting the objection that nullification will foster political anarchy in America because a concurrent majority of states will seldom, if ever, be attained on federal laws. Roman history suggests otherwise.

> To obtain this concurrence, each [order] was compelled to consult the good-will of the other, and to elevate to office, not those only who might have the confidence of the order to which they belonged, but also that of the other. The result was, that men possessing those qualities which would naturally command confidence—moderation, wisdom, justice, and patriotism—were elevated to office; and the weight of their authority and the prudence of their counsel, combined with that spirit of unamity necessarily resulting from the concurring assent of the two orders, furnish the real explanation of the power of the Roman State, and of that extraordinary wisdom, moderation, and firmness which in so remarkable a degree characterized her public men. (Calhoun II:259)[31]

It is obvious to Calhoun that the compromises between such representatives served the common good of both orders. He is confident that in America the operations of the concurrent-majority principle will elevate equally remarkable representatives and enable the federal government to pursue "the common good of all the states" (Calhoun II:255).

The current deficiencies of the union from a republican perspective, then, can be explained as a fall from grace rather than as a natural consequence of its size and diversity. Calhoun can only stand amazed at the prevailing crisis in view of the fact that "our people have been advancing in general intelligence, and, I will add, as great and alarming as has been the advance of political corruption among the mercenary corps who look to Government for support, the morals and virtue of the community at large have been advancing in improvement" (Calhoun II:252–53). No other cause can be assigned for the impending despotism than "a departure from the fundamental principles of the constitution, which has converted the Government into the will of an absolute and irresponsible majority" (Calhoun II:253). In contrast, the special dispensation of South Carolina as a republican community jealous of liberty and suspicious of power can be attributed to its defense of the concurrent-majority principle and its resistance to the absolute-majority principle. For, it, too, "had sunk into avarice, intrigue, and electioneering—from which nothing but some such event [the nullification crisis] could arouse it, or restore those honest and patriotic feelings which had almost disappeared under their baneful influence" (Calhoun II:304).

Furthermore, Calhoun suggests that to the extent the disjunction between South Carolina and the union is natural, it is less because of the difference in size than because of the corruption of power. It is natural for minorities to defend the concurrent-majority principle and liberty and to resist the absolute-majority principle and power, just as it is natural for majorities to act in a converse manner (see Calhoun II:261). It is also natural for majorities to grow, which is the only way Calhoun can account for his state's relative political isolation (see Calhoun II:224). If power corrupts, absolute-majority rule corrupts absolutely.

Calhoun first tests this proposition logically, through a hypothetical demonstration of "the inevitable operations" of the absolute-majority principle. This demonstration proves that the absolute-majority principle reinforces the opposite qualities to the concurrent-majority principle. It rewards narrow, interest-group representation, destroys "every feeling of patriotism," promotes "the most violent party attachment," and caters to "the stronger interest." In sum, it spawns a cycle of consolidationist declension, which regresses through "faction, corruption, anarchy, and despotism" (Calhoun II:248–50).[32]

According to Calhoun, this cycle is rapidly approaching maturity in America. At the end of the dark tunnel of the nation's brief experience with absolute-majority rule is disunion. Calhoun deplores "the deep decay of that brotherly feeling which once existed between these States" that the present controversy "too plainly indicates." He also decries the "madness" of the supporters of the Force Bill, who suppose that the union can be maintained by force once that brotherly feeling has been dissipated. This ominous trend can only be reversed by a return to first principles (see Calhoun II:197, 234–35, 255–56).

For Calhoun, however, a return to first principles is not Publius's bold embrace of a national government. It is the Constitutional Convention's (alleged) decision to retain the essentials of a federal system. In actuality, it is the position of such Anti-Federalists as the Federal Farmer. The paradox of Calhoun's presentation of the concurrent-majority principle as a means of republicanizing the union is that the principle itself is a strongly federal one. This presentation may far surpass anything we saw in the Federal Farmer's case against the Constitution as a blueprint for combining national, federal, and republican features, but it shares the same foundations. It presumes that the individual states are (and always will be) better republics than the union. In Calhoun's view, the states are the natural American republics; the union is a republic, to the extent it is (or can be), through a political artifice. At best, it is a secondhand republic. If the concurrent-majority principle secures a natural aristocracy in the federal government, members of that aristocracy are, in the first place, representatives of their states. If it midwifes a national public good, that good is a composite of state public goods. And if it inculcates patriotism in ordinary citizens, that patriotism is a proxy for primordial state loyalties. The concurrent-majority principle is not part of the Federalist heritage. Publius's extended-republic argument proceeded on quite different grounds. It made size a virtue and placed nature squarely on the side of the union. The American republic was to be an exemplar for its national features, not its federal ones.[33] Indeed, Calhoun's doctrine unfolds as if it were an implicit critique of the extended-republic argument, and from two directions. Not only does size initially raise the problem of regional minorities; size alone cannot solve that problem.[34] The difference between Calhoun's analyses of the two levels of the American polity remains a large one. His acceptance of absolute-majority rule on the state level symbolizes that difference.

We must keep this level-difference in mind as we turn to the question of Calhoun's pluralism. As was true of the Federal Farmer and Publius, his political thought reflects the deep tensions between American nationalism and republicanism. As was true of the Federal Farmer *but not Publius,* his "weak" republicanism on the national level is not so much evidence of pluralism as it is evidence of federal republicanism. Still,

Calhoun does appear to be the most pluralistic of the three statesmen. The attenuation of his republicanism on the state level is the more significant indication of the pluralistic elements in his thought and of the altered relationship between pluralism and republicanism within America's prevailing liberal tradition. By 1833, even federal republicans are relatively pluralistic.

Pluralism and Republicanism

Two dicta to Calhoun's concurrent-majority theory become important to the question of his pluralism.

Near the end of his first nullification speech, Calhoun acknowledges that the states, too, would have to be ruled by concurrent majorities if not for the fact that the federal system relieves them of some of the responsibilities and, especially, revenues of government (see Calhoun II:260). This dictum suggests it is not the special republican qualities of the states but rather the peculiarities of a federal system of government that allows them to be ruled by absolute majorities. As Calhoun's hypothetical demonstration of the operations of the absolute-majority principle shows, the elementary fiscal action of government divides any society, "however small . . . and however homogeneous its interests," into a majority which receives a net gain from public policies and a minority which suffers a net loss (Calhoun II:246). Apparently, the American states also can only be republics by artificial means.

Calhoun's demonstration does prefigure a pluralistic vision of politics. It points to a political vision in which the primary reality is conflict, not consensus; in which the appropriate metaphors are mechanistic, not organic; in which the basic units are egoistic individuals or groups, not cohesive communities; in which the self-assertion of such individuals or groups is the fundamental form of liberty, not their patriotic participation in a larger community; in which their ability to assert their own interests defines their virtue, not their attachment to a greater good; and in which their government representatives exist to expedite that self-assertion, not to integrate them into a richer public life. According to this political vision, a traditional, republican politics, even on a small scale, is not only unrealistic but illiberal. If any utopia is contemplated, it is an anarchic state-of-nature. Calhoun regrets that "the self-government of all the parts" is an ideal "too perfect to be reduced to practice in the present, or any past stage of human society" (Calhoun II:251).

At this point, another one of the dicta to Calhoun's concurrent-majority theory assumes importance. He concludes his hypothetical demonstration of the operations of the absolute-majority principle by calling attention to one difference between how the principle operates in large and in small communities.

> There is, indeed, a distinction between a large and a small community, not affecting the principle, but the violence of the action. In the former, the similarity of the interests of all the parts will limit the oppression from the hostile action of the parts, in a great degree, to the fiscal action of the government merely; but in the large community, spreading over a country of great extent, and having a great diversity of interests, with different kinds of labor, capital, and production, the conflict and oppression will extend . . . [to] a general conflict between the entire interests of conflicting sections, which, if not arrested by the most powerful checks, will terminate in the most oppressive tyranny that can be conceived, or in the destruction of the community itself. (Calhoun II:251–52)

This second dictum suggests a pluralistic argument for smallness, hence for federalism under American conditions. While the game of politics is always violent, it is less violent in smaller, more homogeneous societies. In principle, though—and this is the implication of both dicta—concurrent-majority rule is necessary in all political societies. In this light, Calhoun's solution to the problem of factions does not appear to be so much different from Publius's as merely more complex.[35]

Nonetheless, the pluralist interpretation of Calhoun's concurrent-majority theory is ultimately unsatisfying, just as it was in the case of Publius's extended-republic argument. The pluralist interpretation trades on Calhoun's ambiguity about the basic units of his theory. He refers to individuals, groups, and states (in a federation) in those terms. When he applies the theory to American conditions, however, the American states clearly are the basic units. They are not themselves ruled by concurrent majorities, any more than individuals or groups would rule themselves in such a manner. Is this application of the theory informed by the political realism of a pluralist who realizes that a society composed of sovereign individuals or groups would not be feasible but that a society composed of a limited number of sovereign states might be? Or, is it really informed by the political realism of a republican who realizes that modern societies cannot be ideally small and homogeneous but that the American states come close enough?

For one thing, the pluralist interpretation overlooks the stark contrast between Calhoun's analyses of how the absolute-majority game is being played in the state of South Carolina as opposed to in the federal government. Only the latter seems pluralistic. This contrast suggests that the rules of the game need not be the same for all political societies; more specifically, that concurrent-majority theory is a federal-republican theory in search of properly constructed member-communities which are themselves legitimately ruled by absolute majorities. Calhoun intimates as much when he claims that the ill effects of the absolute-majority game are also being felt in the governments of some of the larger and more populous states (see Calhoun II:260).[36]

Second, the pluralist interpretation neglects the differences between Calhoun's empirical and normative beliefs, not surprisingly, given the general pluralist neglect of such differences. At least as it is being played on the national level and in some of the larger states, the absolute-majority game may be very pluralistic, but that game is hardly Calhoun's ideal. He disdains the logrolling majorities which have enacted prior tariffs.[37] This game is even less acceptable to him than it was to Publius, who also recognized the powerful latent tendencies in any modern, liberal society toward a pluralistic politics. Calhoun certainly seems to think those tendencies are more remediable. He is convinced that a revision in the rules will drastically improve the game, and in an identifiably republican direction. The concurrent-majority game elevates the character of the players and of the whole. Admittedly, Calhoun's faith in this more sophisticated game—the faith of a political scientist in *his* theory—appears somewhat incongruous. Although the desired outcomes are traditional, republican ones, the underlying analysis goes beyond Publius in relaxing the social assumptions which had previously been associated with those outcomes. We should not be startled by this gap in view of the rapid social changes that occurred during Calhoun's historical epoch and of the almost inevitable lag between social and intellectual change. Yet, we remain puzzled by how he got from point "a" to point "b," from a politics of monopolistic factions to a politics of mutual aid (see Calhoun II:224).

In defending a republican interpretation of Calhoun's concurrent-majority theory, I, then, am not denying the very deep tensions in his political thought. At various times, they appear as the tensions between pure and applied theory, political idealism and realism, republicanism and pluralism, and state and national loyalties. The primary contours of his thought, however, were federal-republican ones. Given the powerful factors which opposed such an intellectual synthesis in 1833, the extent to which that seems to have been Calhoun's synthesis is revealing. It not only attests to his ability to develop a coherent political position, but it also manifests the continuing attraction of traditional republican ideas to American statesmen at the dawning of the second-party system. Webster's "ultra"-nationalistic response to Calhoun's nullification doctrine provides even more significant evidence for this thesis.

6

Daniel Webster's Patriotic Community

The comparison of Daniel Webster to Publius is even more natural than the comparison of Calhoun to the Federal Farmer. Webster began his long political career in the early 1800s as a member of the Federalist Party in New Hampshire. He not only invokes the authority of *The Federalist Papers* several times during the course of his major nullification speech, "The Constitution not a Compact between Sovereign States," but the structure of the arguments of the two works is very similar.[1]

Both *The Federalist Papers* and Webster's nullification speech seek to justify a powerful federal or national government in terms of its indispensability to the survival of the union. Webster's principal counterargument—that nullification means disunion—recalls how Publius portrayed the probable consequences of failing to ratify the Constitution. Finally, in their positive arguments, both statesmen stress the value of the union.

Their arguments, however, diverged in several important respects. In the first place, Webster could appeal to a legal tradition with strong nationalist precedents, including *The Federalist Papers* as well as a number of Supreme Court cases he himself had argued before the bench. This tradition obviously had not been available to Publius. Yet, this factor was not conclusive, in part because Calhoun could marshall his own states-rights precedents.[2] At a deeper level, Webster's differences with Publius flowed from his focus on "the Constitution not a compact," which was less a focus on the legality of the union or the value of the union, or even on its sociology, than on its historicity. Again, this line of argument had not been open to Publius, at least not as open to him. This factor seemed more conclusive than the first because Calhoun found it impossible to deny all the prerequisites of union.

The contrasting slants Publius and Webster took on the union were significant insofar as they reflected the more pluralistic character of Webster's nationalism. Those slants, then, also reflected the way American political ideas and social conditions had evolved since 1787–88 and the way Webster was struggling to adjust the one to the other. In this effort, he was able to take advantage of a third post-Constitution development. This was the emergence of a distinctive national-republican tradition, a tradition which *The Federalist Papers* also had forwarded and which Webster could now claim had superseded the older federal-republican tradition to which Calhoun appealed. He, thus, could present himself as a traditionalist, in this respect more like the Federal Farmer than Publius.[3]

Otherwise, there are few traces of Anti-Federalism in Webster. His speech is extraordinary for how little it concedes to the other side, despite the cogent political and rhetorical reasons to temper his ultra-nationalistic understanding of the Constitution which were so evident in *The Federalist Papers*.

In a sense, Webster was more isolated than Calhoun was. His claim that nullification was patently unconstitutional was widely accepted. The path he took to that result, however, was quite different from the one other senators took. Most of them started from states-rights premises, distinguishing themselves from Calhoun by asserting the bindingness of a constitutional compact between (now less) sovereign states.[4] Although Webster could, and did, stoke the nationalist sentiments the crisis had aroused, those sentiments did not necessarily translate into support for his ultra-nationalistic interpretation of the Constitution. He, furthermore, was in a distinct minority in opposing a compromise tariff.[5]

All in all, the standard view that Webster's speech was an ambitious bid for rapprochement with Jackson is unconvincing. He would have threatened South Carolina with a bigger "stick" than the president did, and he would not even have offered it a "carrot." If he had wanted to move toward Jackson and the political middle on this set of constitutional and economic issues, he adopted a peculiarly awkward course.[6]

Webster's inflexibility on the tariff calls attention to the other prevailing view of his speech. This view emphasizes his faithful representation of New England's economic interests. At least heretofore, his interpretation of the Constitution had shifted in tandem with those interests.[7]

Nevertheless, we should not overlook how Webster, like Calhoun, was locking himself into a particular constitutional position which would make future shifts more difficult. In this speech, he argues that the Constitution not only permits Congress to enact protective tariffs but positively enjoins it to do so (see Webster VI:228–32). He does *not* make the more politically malleable argument so prominent in his other (post-1828) speeches; namely, that the economic interests protected by high tariffs are really national, not sectional, ones.[8]

In their nullification speeches, both Webster and Calhoun largely bypass the tariff issue for what they consider the deeper constitutional and metaconstitutional issues involved in the controversy.[9] This intention does not prove that their positions were uncontaminated by their constituents' economic interests or their own political ambitions. It, however, does suggest that we cannot adequately understand their positions in terms of only those factors. Or, to make a stronger claim, it suggests we can construct a meaningful narrative of the nullification debate which ignores those factors.[10]

Calhoun's apposition of liberty and union to capture his fundamental differences with Webster is a half-truth. Webster is the staunch defender of the legal and historical union. Yet, he is also the architect of "liberty and union," though in a much different sense than Calhoun ever could claim to be. Webster believes that the union is intimately connected to American liberty as a national, not a federal, republic.

In the three sections of this chapter, I will explore the basic components of Webster's refutation of nullification doctrine—nationalism, republicanism, and the legitimation of the first by the second. His speech will be analyzed into: (1) a constitutional argument for the legal supremacy of the government of the union; (2) an empirical argument for the republican character of the union and its government; and (3) a normative argument for the value of the union. In contrast to existing Webster scholarship, I will show how his nationalism, which undoubtedly was the most salient feature of his mature political thought, was a national republicanism.[11]

National Government: The Legal Tradition

Webster contends that the Constitution is a fundamental law which establishes the primary rules of the American system of government (see Webster VI:198–99). His initial argument is self-evident. If the Constitution establishes the primary rules of the American system of government, then the proper method for settling disputes, such as the present one, over the nature of that system of government is to refer to the text of the Constitution. The proper method is *not* to engage in "philosophical remark upon the general nature of political liberty, and the history of free institutions; and upon other topics, so general in their nature as to possess, in my opinion, only a remote bearing on the immediate subject of this debate" (Webster VI:184)—which is what he accuses Calhoun of doing. He intends to hold Calhoun "to the written record. In the discussion of a constitutional question, I intend to impose upon him the restraints of constitutional language" (Webster VI:190). His initial criticism of Calhoun's Senate resolutions is also self-evident. They do not use constitutional language. Webster admits that if the Constitution declared itself a compact between sovereign states, then he would have to yield to the senator from South Carolina. Using constitutional language, however, there is "a wide and awful hiatus between his premises and his conclusions" (Webster VI:188).

According to Webster, the Constitution calls itself a constitution, not a compact between sovereign states. It also calls itself the supreme law of the land and designates the Supreme Court as its final arbiter. Webster insists that "friends and foes" of the Constitution agreed that it provided for a federal government that would be the judge of its own powers instead of subjecting those powers to state governments or conventions, at least not to less than three-fourths of them (see Webster VI:197–99, 215–19, 222).[12]

On the basis of constitutional language, Webster concludes that nullification (like secession) can only be a revolutionary right. The Constitutional Convention's decision to reject a congressional veto over state laws was a implicit decision for judicial review, not for nullification. He claims that this is how Publius presented the matter (see Webster VI:217).[13] The framers of the Constitution would surely have realized that a plan of government which countenanced nullification would not have materially improved on the Articles of Confederation. Even in 1833, nullification would revisit Americans with the very miseries the founders were trying to escape through a new constitution. Webster foresees still more horrible consequences. The alarming scenarios include war between the federal and South Carolina governments, war between the states, and, worst of all, disunion. These, again, were the very possibilities the founders were seeking to avoid through the Constitution (see Webster VI:192–95, 211, 223–26).

If these evils occur, though, they should not be blamed solely on the nullifiers. Opponents who share a compact view of the Constitution must shoulder some of the blame. Webster also credits the founders, and specifically Publius, with being cognizant of the inadequacies of such a view of the Constitution.

> Its most distinguished advocates, who had been themselves members of the Convention, declared that the very object of submitting the Constitution to the people was, to preclude the possibility of it being regarded as a mere compact. "However gross a heresy," say the writers of the Federalist, "it may be to maintain that a party to a compact has a right to revoke that compact, the doctrine itself has had respectable advocates." (Webster VI:209)[14]

Webster's reference to Senator Rives's speech must be understood in this light. He states that Rives's opinions "are redolent of the doctrines of a very distinguished school, for which I have the highest regard, of whose doctrines I can say, what I can also say of the gentleman's speech, that, while I concur in the results, I must be permitted to hesitate about some of the premises" (Webster VI:200).[15] The dilemma is that while Rives's results differ from Calhoun's, his premises do not. Webster not only questions the validity of the premises but doubts whether anyone who adopts them can consistently refuse to accept Calhoun's results.

Rives had argued that the Constitution is a compact between sovereign states, yet that it is binding on any one of them. No one party to the constitutional compact can authoritatively interpret its rules. South Carolina, therefore, cannot, even contingently, suspend the force of a federal law on grounds of unconstitutionality. Only the states collectively can perform such an action through the amending process (see *Debates* IX:500–501).[16] Rives saw himself as exposing the heresies of nullification doctrine. Webster assumed the same role. He, nevertheless, could not wholeheartedly praise Rives's efforts. Webster's insight, or, rather, the insight of Webster's Publius, was that if the Constitution is understood as a compact between sovereign states, it will always be a matter of controversy whether one state, or a minority of states, may revoke the terms of that compact. (Here, Webster was prophetic. It was a matter of controversy in America until it exploded into the secession crisis and civil war three decades later.[17])

One obvious way of bolstering the case against nullification would be to start from different premises. Instead of challenging Calhoun on his own states-rights premises, Senators Frelinghuysen and Clayton asserted the nationalist premise that the Constitution was a compact between the American people as a whole which formed a sovereign, national community. They, then, defined the Constitution as a national or social compact, not a federal one (see *Debates* IX:313–16 [Frelinghuysen]; 387–88 [Clayton]).

Webster, however, does not take this approach. Although he views the union as a sovereign, national community, he does not consider the Constitution constitutive of that community. He seems to have in mind both the historical accuracy and the political adequacy of different beliefs about the nature of the Constitution and the union. The Constitution not only was not the formative compact of the union, but Americans should not think of it in such terms because that belief robs the union of its psychological permanency. Whether that was Publius's deeper understanding or not, Webster borrows his authority to suggest the historical inaccuracy and political inadequacy of *all* compact views of the Constitution. Where other senators debated whether the Constitution was a compact between sovereign states, Webster insists that it is not a compact at all.[18]

It is true that Webster speaks of the Constitution as a social compact. He must explain the compact language some of the state conventions used in ratifying the document.

> The consent of the people has been called, by European writers, the social compact; and, in conformity to this common mode of expression, these conventions speak of that assent, on which the new Constitution was to rest, as an explicit and solemn compact, not which the states had entered into with each other, but which the people of the United States had entered into. (Webster VI:210)[19]

Yet, Webster's first point is that the Constitution is *not* a federal compact. And his second point is that it is only loosely speaking a social compact. That usage is really a misapplication of the language of "European writers" to American events. It is on this basis that he elaborates his differences with Rives.

> I do not agree that the Constitution is a compact between States in their sovereign capacities. I do not agree, that, in strictness of language, it is a compact at all. But I do agree that it is founded on consent or agreement, or on compact, if the gentleman prefers that word, and means no more by it than voluntary consent or agreement. The Constitution, Sir, is not a contract, but the result of a contract; meaning by contract no more than assent. Founded on consent, it is a government proper. (Webster VI:200–201)

Webster's meaning is somewhat obscure. But the suggestion is that the mere fact the American people consented to the Constitution does not make it a social compact, any more than a successful popular referendum on an ordinary law would have that effect. Webster claims that anyone who attempts to describe the events surrounding the ratification of the United States Constitution employing compact language is bound to experience great difficulties because the language just does not fit the events. He earlier had assailed the "phraseology" of Calhoun's resolutions in these terms.

> . . . this phraseology tends to keep out of sight the just view of a previous political history, as well as to suggest wrong ideas as to what was actually done when the present Constitution was agreed to. In 1789, and before this Constitution was adopted, the United States had already been in a union, more or less close, for fifteen years. At least as far back as the meeting of the first Congress, in 1774, they had been in some measure, and for national purposes, united together. Before the Articles of Confederation of 1781, they had declared independence jointly, and had carried on the war jointly, both by sea and land; and this not as separate States, but as one people. (Webster VI:187)[20]

For Webster, then, the compact view of the Constitution clearly is historically inaccurate. The Constitution did not make the American people one people. The American people already had formed a union. What was "founded" in 1787–89 was "only" a new government.

Like other Senate critics of South Carolina's actions, Webster also ascribes a binding character to the Constitution.

> The Constitution, Sir, regards itself as perpetual and immortal. It seeks to establish a union among the people of the States, which shall last through all time . . . The instrument contains ample provisions for its amendment, at all times; none for its abandonment at any time. It declares that new States may come into the Union, but it does not declare that old States may go out. The Union is not a temporary partnership of States. It is the association of the people, under a constitution of government . . . This instrument [is] to be regarded as a permanent constitution of government. (Webster VI:211)[21]

However, Webster, unlike the other critics, considers the Constitution binding in this strong sense precisely because it is not a compact. Similarly, he thinks that a union, which is not based on a compact, is intended to be permanent while a league, which is based on a compact, is not. He distinguishes the Constitution and the Articles of Confederation on these grounds. On the one hand, the Constitution did not declare itself perpetual because it only established a new government, which has the expectation of perpetuity. On the other hand, the Articles, though it only established a league of state governments, declared itself perpetual because the American union was already more than a league (see Webster VI: 187–88, 190–91).

Besides "social compact," Webster tries to discourage the use of another key word in European political grammar. He contends that American political discussions are also frequently confused by references to the federal division of powers as dual or divided "sovereignty."

> The sovereignty of government is an idea belonging to the other side of the Atlantic. No such thing is known in North America . . . with us, all power is with the people. They alone are sovereign; and they erect what governments they please, and confer on them such powers as they please . . . we only perplex ourselves when we attempt to explain the relations existing between the general government and the several State governments, according to those ideas of sovereignty which prevail under systems essentially different from our own. (Webster VI:202)

It, therefore, was the sovereign people in America which divided powers between the governments on the two levels of the nation's peculiar federal system. When speaking most precisely, all the major figures in the nullification debate agreed that, of course, peoples, not governments, are sovereign and that, yes, logically sovereignty is indivisible—which was one of Calhoun's *idée fixe* (see Calhoun II:231–32).[22] They disagreed on three further questions: (1) which American people is sovereign, the people as a whole or as divided into states: (2) in that relation, is there something analogous to divided (popular) sovereignty; and (3) in the relation between the federal and state governments, is there not also something analogous to divided (government) sovereignty.

According to Webster, the preamble is presumptive evidence that it was the American people as a whole which ordained that the relationship between the federal and state governments would be one of hierarchy rather than co-equality (see Webster VI:210).[23] With respect to neither popular nor government "sovereignty," though, did the Constitution emasculate the states. Federal supremacy does not mean that the state governments are not governments with their own wills to act directly on their own citizens, nor does national sovereignty mean that the peoples of the states are not peoples with their own wills to make their own constitutions and governments. In fact, the initial impetus of Webster's attack on Calhoun's resolutions is to show that the American political system has this dual aspect as against a unimodal state sovereignty (see Webster VI:199, 209–11). Still, that demonstration is but a resting place in the process of showing that in this dual aspect the relationship between union and states is closer to that between states and townships than federations and member-states.

As we have seen, Calhoun claimed this was exactly how his opponents were falsely portraying the relationship. He took the other extreme to Webster. Most of the other participants in the debate took more moderate, "partly federal, partly national" or "divided sovereignty" positions.[24] Calhoun's theory of indivisible sovereignty was clearly directed at them, attempting to force them to seize one of the extremes. Yet, all theories of sovereignty did, as Webster observed, only perplex when applied to the American case. They distorted what, in reality, was a very complex political system (see Webster VI:210–11).

Although Webster does not explicitly call Calhoun's theory "metaphysical," as Clayton did, his critique is similar. He finds that Calhoun's many abuses of American political grammar combine to create a highly misleading impression of the nature of the Constitution (see Webster VI:185–87). Calhoun's jargon, however, is only a sign of the more fundamental flaws in nullification doctrine. Webster continually expresses astonishment at its pure presumptiveness. It "has nothing to stand on but theory and assumption." "In place of plain historical facts" it substitutes "a series of assumptions." Is it not "the mere effect of a theoretical and artificial mode of reasoning upon the subject? a mode of reasoning which disregards plain facts for the sake of hypothesis?" (Webster VI:190, 202, 213).

Like Dickens's Thomas Gradgrind in *Hard Times,* Webster appears to want nothing but the facts. In his nullification speech, he adopts the cast of mind of a grammar-school teacher who constantly derides the idle fancies of his wildest-eyed pupil; or, equally, of a lawyer who repeatedly objects to his adversary's efforts to introduce mere theory and assumption into the case. But perhaps the best analogy is to Emerson and Publius. Webster is the American scholar, defying those who would search beyond American historical experience for their political norms. What that experience is, though, has become no more unequivocal through time, and probably less so. Calhoun can also dismiss Webster's ultra-nationalistic language as "not very American" (Calhoun II:294). Webster, thus, must go beyond plain facts and constitutional language to refute nullification doctrine. Even if he does eschew Calhoun's more philosophic tangents, his stance is not simply a historical one.[25]

Popular Government: The Republican Tradition

Webster's methodological critique of Calhoun's Senate resolutions is not limited to their deficiencies in American political grammar. Calhoun's "theoretical and artificial mode of reasoning" is also contrasted with "the most just and philosophic view" of the American founding. The most philosophic view happens to be "the plainest account" not only because it is the most historically accurate but because it is the most consonant with the nature of popular government (Webster VI:201–2).

Among the many things nullification doctrine denies, it denies "the practicability of all free governments." Webster wonders whether Calhoun really means to suggest that the founders labored for three months to establish a free government which is, in practice, unworkable or, alternatively, despotic. In the course of impugning this heresy of nullification doctrine, Webster investigates the assumptions of free or popular government. Not surprisingly, those assumptions turn out to be two of the major assumptions of republican thought: the existence of an overarching public good and the prevalence of civic virtue among the general populace.

> Mr. President, all popular governments rest on two principles, or two assumptions:—
>
> First, that there is so far a common interest among those over whom

> the government extends, as that it may provide for the defence, protection, and good government of the whole, without injustice or oppression to parts: and
>
> Secondly, that the representatives of the people, and especially the people themselves, are secure against general corruption, and may be trusted, therefore, with the exercise of power. (Webster VI:222)

Webster implicitly argues that according to these two assumptions of popular government the union can sustain—who, beside Calhoun, would deny that it is sustaining?—a popular government and that it, therefore, is a national republic. Indeed, this portrait of the union provides the subtext of his speech, just as it did of *The Federalist Papers*. These assumptions also constitute a thinly veiled critique of Calhoun's doctrine. Webster seeks to prove that Calhoun is not a republican for denying that the union can fulfill the assumptions, when that merely proves he is not a national republican. Moreover, Calhoun's position is strengthened to the extent that Webster can only demonstrate that the union fulfills the assumptions in a weak sense.

Interests and Sentiments

In reference to the first assumption of popular government, Webster does not really discuss a common *interest*. This "failure" is striking given that the occasion seems to demand such a discussion. The nullified tariffs seem to be victims of the apparent lack of a common interest in the union as a whole. Webster's only acknowledgement of that widely shared perception is the assertion that nullification "appeals from the common interest to a particular interest" (Webster VI:218).[26] Mostly, he discusses a common *feeling*.

Both at the beginning and the end of his speech, Webster predicts that the union as a political sentiment will suppress nullification. After first investing his own trust in public opinion, he metaphorically denies Calhoun the same opportunity. Webster claims that "the cause which he [Calhoun] has espoused finds no basis in the Constitution, no succor from public sympathy, no cheering from a patriotic community" (Webster VI:182). In the long peroration of his speech, Webster returns to this theme, again implicitly contrasting his own situation with Calhoun's.

> Be assured, Sir, be assured, that, among the political sentiments of the people, the love of union is still uppermost. They will stand fast by the Constitution, and those who defend it. I rely on no temporary expedients, no political combination; but I rely on the true American feeling, the genuine patriotism of the people, and the imperative decision of the public voice. (Webster VI:237)

Undeniably, most of what Webster says in between these two passages is calculated to demonstrate that nullification doctrine "finds no basis in the Constitution." Through the organization of the speech, he, nonetheless, stresses his belief that the doctrine's falsity ultimately depends on its rejection by public opinion. He is also confident of success by that standard because "love of Union is still uppermost."[27]

This emphasis on public opinion should not be understood as simply a recognition of the power of public opinion in republican societies. Nor should it be understood as simply part of a rhetorical strategy of appealing from one presumed state of public opinion to another: from local to national loyalties. It is those things. However, it is also an integral part of Webster's portrait of the union as a national republic. He centrally defines the union as "a patriotic community."

No matter how persuasive this definition of the union may or may not be, it is a nontraditional one. Webster departs from the republican tradition in fitting the union to the first assumption of popular government. He does not discuss a common interest or public good but the social homogeneity which lies behind it. Then, in discussing social homogeneity, he discusses ideological homogeneity, which does not as clearly lead back to a common interest as the economic homogeneity he fails to discuss does.

Of course, economic and ideological homogeneity are not mutually exclusive choices. Nor can they be sharply distinguished as traditional and nontraditional emphases. Webster does not discard the assumption of an union of interests. His departure, instead, consists in placing added weight on a union of sentiments, implying that republics rest on common feelings more than common interests. This emphasis is the more realistic one given: (1) the prevailing social conditions in America; (2) the probable social conditions in any modern nation, especially large ones; and (3) the specific conflict of interest which provides the immediate context of his speech. While it can be viewed, negatively, as a response to increasing economic complexity, it can also be viewed, more positively, as a response to the steady growth of popular nationalism in America. Webster appears to be extending Publius's intellectual synthesis of nationalism and republicanism in the only way he realistically can extend it.[28]

Dialectically, Webster's definition of the union as a sentiment emerges as a response to the challenge posed by Calhoun's strongly economic definition of a federal union. Whether Webster thinks he can or cannot meet that challenge on its own grounds, he chooses to join issue on the level of Americans' divided loyalties. There, he professes to be certain of victory. In this light, the worst sin of nullification is the pathetic image it projects of the union; that it is not the uppermost sentiment in America, particularly not in South Carolina. Webster considers that image a false one. If allowed to stand, it, nonetheless, would shake the foundations not only of the real union but of his own intellectual synthesis of nationalism and republicanism.

Before turning to Webster's rebuttal of the ideological implications of nullification doctrine, which concerns the second assumption of popular government more than the first, we must explore several corollaries to his definition of the union as a sentiment. This definition dictates fairly specific answers to the questions of how the union came into being and of how to structure its system of government. These deductions become major features of Webster's political argument.

For Webster, an integral part of the national-republican position is that the union, over and above the states, was formed in an organic manner. This empirical premise follows from his beliefs about the nature of the union. Since it is essentially a sentiment, it could not have been formed by rational agreement. His implicit metaphor for the union is a marriage, where (ideally) the compact only recognizes what already exists.

As we have seen, the way Webster describes the union coming into being is dissociated not just from the Constitution but from a compact of any kind. It seems to have been the unintended product of the growing together of a patriotic community. The "birth" of the union cannot even be precisely dated.

In Webster's historical narrative, the union existed as a sentiment before American independence was declared. War and the achievement of independence greatly strengthened this "pre-existing" union. As prefigured in the preamble, the Constitution accelerated the process of perfecting the union. The Articles of Confederation had not. The prospect of a state nullifying a federal peace treaty—a painful reminder of Confederation embarrassments—provokes Webster to exclaim:

> The truth is, Mr. President, and no ingenuity of argument, no subtilty [*sic*] of distinction can evade it, that, as to certain purposes, the people of the United States are one people. They are one in making war, and one in making peace; they are one in regulating commerce, and one in laying duties of impost. The very end and purpose of the Constitution was, to make them one people in these particulars; and it has effectually accomplished its object. (Webster VI:205)

This passage elucidates Webster's beliefs about the nature of the union. The American people are one with respect to war, peace, and commerce. The passage, however, should not be read as stating a claim that the Constitution, primordially as a social compact, made the American people one in those respects. The claim is rather that the Constitution gave explicit legal definition to prior feelings of oneness. Furthermore, the passage does not necessarily express a commitment to federalism, as if the Constitution were designed to unite the American people into one nation for some purposes and into various states for other purposes through a specific division of government powers. Two considerations oppose this reading. First, Webster thinks that no people is one with respect to more than a few purposes under any popular government. This is due to the limited nature of such governments (see Webster VI:222). Second, he thinks that the Constitution placed the highest common purposes of any civilized people under federal supervision. For the first time in American history, it matched energetic federal powers with enduring national purposes.[29] It could not help but boost feelings of national unity and perfect the union in this crucial, extralegal sense. In this sense, the Constitution did much more than merely establish a new government.

Webster insists on the superiority of federal powers under the Constitution and, more importantly, on their unifying effects.

> This government may punish individuals for treason, and all other crimes in the code, when committed against the United States. It has power, also, to tax individuals, in any mode, and to any extent; and it possesses the further powers of demanding from individuals military service . . . No closer relations can exist between individuals and any government. On the other hand, the government owes high and solemn duties to every citizen of the country. It is bound to protect him in his most important rights and interests. It makes war for his protection, and no other government in the country can make war. It makes peace for his protection, and no other government can make peace . . . He goes abroad beneath its flag, and carries over all the earth a national character imparted to him by this government and which no other government can impart. In whatever relates to war, to peace, to commerce, he knows no other government. All these, Sir, are connections as dear and as sacred as can bind individuals to any government on earth. (Webster VI:203)

In fact, Webster does not mention the residuary powers of the state governments, despite his explicit endorsement of a federal division of powers and despite the rhetorical advantages to be gained by reiterating how powerful the state governments seem to remain under the Constitution (even if they cannot nullify federal laws). After all, the administration of justice and taxation, unlike war, peace, and (foreign) commerce, are concurrent powers. Webster's *modus operandi,* though, is to constantly remind his audience of the inferiority of the state governments, both as to the objects under their care and the supremacy of federal to state law. These reminders are hardly conducive to feelings of state pride.[30]

Webster's rhetoric has a different structure than Publius's, in part because of their different polemical situations, in part because of the broader differences in social context, and in part just because they were different intellects. Publius was much more solicitous of localist sentiments. They were relatively stronger in 1787–88, and the Constitution's prospects were uncertain largely because of them. Publius linked the survival of the union to a particular system of government. He did not, at least not centrally, link it to a particular configuration of public opinion. He focused much more on such palpable causes of disunion as foreign intrigues, commercial rivalries, and border disputes. To Webster, the first cause of disunion is the erosion of patriotic sentiments. His rhetoric is primarily intended to cultivate those sentiments, not to convince his audience of the validity of a certain set of political propositions. He seems convinced that any slippage in those sentiments could be fatal to the union.[31] Yet, the area of agreement between the two statesmen is still substantial. Both stress the importance of federal supremacy to continued union; both assume that Americans cannot long remain one people without one government possessing powers to directly serve their highest common purposes; and, finally, both see the preservation of the union as being contingent on their own political success.

Webster's equation of Calhoun's resolutions to the Articles of Confederation is compelling because both "documents" deny federal supremacy. Notwithstanding all Calhoun's protestations to the contrary, Webster claims that nullification can only mean disunion.

> It strikes a deadly blow at the vital principle of the whole Union. To allow State resistance to the laws of Congress to be rightful and proper, to admit nullification in some States, and yet not expect to see a dismemberment of the entire government, appears to me the wildest illusion, and the most extravagant folly. The gentleman [Calhoun] seems not conscious of the direction or the rapidity of his own course. The current of his opinions sweeps him along, he knows not whither. To begin with nullification, with the avowed intent, nevertheless, not to proceed to secession, dismemberment, and general revolution, is as if one were to take the plunge of Niagara, and cry out that he would stop half way down. (Webster VI:193–94)

The moral of Webster's speech is that to the extent Americans value the union they must condemn nullification. His frequent references to the founders are not only intended to help explicate the meaning of the Constitution but to identify the goals behind the Constitution as goals which are still highly desirable and not yet entirely secure.

Calhoun, however, rejects, as the Federal Farmer did, the supposition that federal supremacy is a necessary condition of continued union. He strongly challenges Webster on that premise. The nature of their dispute, though, is seriously complicated by their different definitions of the union. It is less clear that a lack of federal supremacy would destroy a federal republic than that it would destroy a national republic.

Far from nullification doctrine proving that Calhoun was not a republican, it proves that he was a federal republican. He did not repudiate Webster's first assumption of popular government on the state level, neither in terms of common interests nor common feelings. In a federal republic, citizens should be—and Calhoun undoubtedly believed American citizens were—united in their economic interests and patriotic sentiments more as divided into states than as a whole. Only Webster's failure to distinguish Calhoun's positions on the federal and state levels makes him appear nonrepublican. His vulnerable point was not any inconsistency between nullification doctrine and republicanism but between nullification doctrine and nationalism. It was the fact that he himself wanted the union to be more than a traditional federal republic. Webster's nationalistic appeals to a patriotic community were the most harmful to Calhoun's cause since so many Americans, including Calhoun himself, were sensitive to such appeals. His predicament was very similar to the one in which the Federalists had put the Anti-Federalists. The national-republican position, though, was not without its own weaknesses.

The principal measure of national homogeneity Webster invokes is Americans' common attachment to the union. This measure involves an attenuation of the degree of social homogeneity which had traditionally been thought necessary in republics. It also involves a shift from economic to ideological homogeneity. Webster's position here is even more pluralistic than Publius's had been. He does not assert that Americans are "a people descended from the same ancestors, speaking the same language, professing the same religion" (Publius, p. 37, paper 2). He cannot assert that with the same credibility Publius could in 1787. Webster's greater pluralism reflects the dramatic social changes which have occurred in the interim. He, of course, would have been chiefly concerned with how those changes might negatively affect the persuasiveness of his nationalist position, not with how they might lead him to dilute his republicanism. Nevertheless, we should not overlook his refusal to be more realistic (persuasive?) in either of two ways: (1) by moving backward in Calhoun's federal-republican direction and the more pronounced social homogeneity on the state level; (2) by moving further forward in a pluralistic direction and the prescription of an even less tightly knit society on the national level. These refusals delimit his national republicanism.[32]

Now, Webster did not decide by fiat, or uniquely, or even self-consciously, to accept looser measures of social homogeneity. His acceptance was not only a natural reaction to social change; the whole American liberal tradition had been evolving toward a similar, more pluralistic point of view.[33] In this sense, Webster was merely exploiting an intellectual trend which was favorable to the nationalist position. He, however, was also building on Publius's national-republican tradition, a relatively new (sub)tradition that allowed for looser measures of social homogeneity without becoming completely pluralistic. Above all else, he could legitimately claim to be following *that* tradition.

Webster's investigation of the second assumption of popular government further elaborates the in-betweeness of his political position and yet how it remained more republican than pluralist.

Sentiments and Virtues

The second assumption of popular government professes faith "especially" in the civic virtue of the people. Webster's argument for how Americans fulfill the first assumption of popular government is his argument for how they fulfill the second. If a patriotic *community* is primarily how they share a common "interest," then a *patriotic* community is primarily how they are secure against general corruption. Once more echoing Publius, he contends that Americans are good citizens not in terms of civic virtue but in terms of patriotism. In a way which far surpasses Publius, though, he seems to believe that the union is a republic just because it is a patriotic community.

The anomaly in Webster's invocation of a patriotic community is that it has not heretofore asserted itself in South Carolina. If the South Carolina unionists tarnish the republican imprint on Calhoun's image of his own state, then the nullifiers similarly affect Webster's image of the union. Webster metaphorically purges the union of its alien elements in a long passage near the end of his speech.

> It is, Sir, only within a few years that Carolina has denied the constitutionality of these protective [tariff] laws. The gentleman [Calhoun] himself has narrated to us the true history of her proceedings on this point. He says, that, after the passing of the laws of 1828, despairing then of being able to abolish the system of protection, political men went forth among the people, and set up the doctrine that the system was unconstitutional . . . I deem it a great misfortune, that, to the present moment, a great portion of the people of the State have never yet seen more than one side of the argument. I believe that thousands of honest men are involved in

> scenes now passing, led away by one-sided views of the question, and following their leaders by the impulses of an unlimited confidence. Depend upon it, Sir, if we can avoid the shock of arms, a day for reconsideration and reflection will come; truth and reason will act with their accustomed force, and the public opinion of South Carolina will be restored to its usual constitutional and patriotic tone. (Webster VI:236)

This passage is remarkable not only for how it subverts the parallel passage in Calhoun's first nullification speech on the political situation in South Carolina. It is also remarkable for how it substitutes patriotism for civic virtue in a standard republican argument. The basic patriotism of the American people is incorruptible; if some Americans are not acting patriotically at any point in time, it is because they have been misled by a faction of disingenuous politicians; given more time and information, their basic patriotism will reassert itself.

Webster's substitution evidences movement toward a more pluralistic conception of citizenship to the extent that his idea of patriotism is an attenuation of traditional republican ideas of civic virtue. Yet, it is precisely because he could substitute patriotism for civic virtue without doing much violence to the republican tradition that he remained a republican statesman.

The substitution works because patriotism, like civic virtue, is a sentiment directed toward the whole. Insofar as patriotism is the uppermost sentiment in Webster's American creed, he downgrades the other, more self-directed sentiments underwritten by pluralists. Patriotism is also similar to civic virtue in only having meaning within a community. There must be a patriotic community which attracts the sentiments of its members and for which they are willing to make some significant sacrifices. According to pluralists, the existence of such a community is both improbable and unnecessary.

Patriotism, though, is not civic virtue. It does not demand the same active participation in politics. It is generally a quiescent sentiment, constantly nurtured but only called into action on extraordinary occasions, such as Webster considers the present one. Even then, it is not at all clear what he is asking his American patriots to do. The citizens of South Carolina are being urged to act in the most concrete ways. They should renounce nullification and yield what many of them firmly believe to be their true economic interests in order to maintain a patriotic community. Nonetheless, Webster himself believes that their sacrifice is only an apparent one. It is based on the fallacy that protective tariffs are (unconstitutional) bounties on their labor. Webster brushes lightly over this point in his speech because he wants the debate to turn on sentiments, not interests or principles. This tactic emphasizes his conviction that the nullification crisis concerns how Americans should feel more than how they should act. He does not even recommend a specific course of action to his immediate audience in the Senate.[34]

Perhaps, Webster substitutes patriotism for civic virtue because he thinks most Americans do not possess sufficient participatory skills, inclinations, or opportunities to practice civic virtue, particularly in the context of national political controversies. Or, his substitution may be informed by the realization that no citizen body, however small and homogeneous, possesses those qualities. His more pluralistic conception of citizenship may be essentially defined by either his nationalism or his social realism. It probably has roots in both. Still, it retains strong affinities to traditional republican conceptions of citizenship, thus suggesting that his political thought was also influenced by an identifiably republican tradition of ideas. After all, the ambiguity of South Carolina's sacrifice on the tariff can, at least partially, be attributed to the ambiguity intrinsic to traditional republican views of the relationship between the public good and civic virtue.

Webster's gloss on the South Carolina nullification movement more clearly suggests a republican influence in other respects. In the first place, it discloses his anti-party bias. He denounces the nullifiers as a faction of "political men" who have offered the people of South Carolina only "one-sided views of the question." He covers their whole activity with the broad brushstroke of illegitimacy. He himself shuns such conduct. He disclaims any reliance on a "political combination" in presenting his own views to the American public because he is merely appealing to its "genuine patriotism" (Webster VI:236–37).[35]

In the second place, this passage focuses attention on the person of Webster, in his role as a republican statesman. The people of South Carolina have been "led away" by political propaganda and "impulses of an unlimited confidence" in their political leaders. Not time alone will restore public opinion in the state to "its usual constitutional and patriotic tone." Someone like Webster must act as a catalyst. In the peroration of his speech, he vows to stand "among those upon whom blows may fall first and fall thickest . . . to prevent the Constitution from being nullified" and he calls on "the people to come to its [his?] rescue" (Webster VI:238).[36] He more self-consciously reflects on his role as a republican statesman at the opposite end of the speech, in his allusion to the Hayne debate.

> There has been a time, when, rising in this place, on the same question, I felt, I must confess, that something for good or evil to the Constitution

> might depend on an effort of mine. But circumstances are changed. Since that day, Sir, the public opinion has become awakened to this great question; it has grasped it; it has reasoned upon it, as becomes an intelligent and patriotic community, and has settled it, or now seems in the progress of settling it, by an authority which none can disobey, the authority of the people themselves. (Webster VI:183)[37]

Webster's role, though, differs depending on whether he is appealing to a patriotic or an intelligent community. He only has to signal that the union is in danger to arouse the patriotic community in America. His repeated claims that nullification means disunion define his appeal to that community. In appealing to the intelligent community in America, he faces the somewhat more arduous task of demonstrating that nullification is unconstitutional. An intelligent community presumably is one that can correctly judge the truth of doctrines by their appropriate standards; in this instance, nullification doctrine by the Constitution.

While an intelligent community is disposed to the truth, its relation to the truth is tentative. Americans have "reasoned" on the "great question" raised by the nullification controversy. Yet, it was necessary for a Webster to guide that process. Three years after the Hayne debate, the people of South Carolina, part of this intelligent community, still exist in a false relation to the truth. Webster is convinced that momentarily "truth and reason will act with their accustomed force" in South Carolina, but, again, it is incumbent on someone like him to combat the spurious doctrines of her counterfeit leaders.

Webster's references to his own relationship to his audience recall the traditional republican view of the relationship between natural aristocrats and the broader community. Implicitly, Webster uses his own relationship to his audience to demonstrate that the union satisfies the second assumption of popular government in terms of the general incorruptibility of both the representatives of the people and the people themselves. Not coincidentally, he presents that relationship as far superior to the one which prevails on the state level, as in South Carolina between Calhoun and his misguided followers. Webster contends that the federal government has proven less susceptible to factions, just as Publius had predicted. For Webster, as for Publius, this superiority is one of the main stays of the national-republican position.

As already noted, the elocutionary force of Webster's two assumptions of popular government is that Calhoun, in rejecting those assumptions, is not a republican statesman. Webster claims that Calhoun rejects the second assumption because his concurrent-majority principle would unduly restrict majority rule. At bottom, Calhoun's theory shows a lack of faith in the virtue of the American people and of their elected representatives, or so Webster argues.

According to Webster, if Calhoun's theory were put into practice, "then a minority, and that a small one, governs the whole country." Webster continues:

> Sir, those who espouse the doctrines of nullification reject, as it seems to me, the first great principle of all republican liberty; that is, that the majority must govern. In matters of common concern, the judgment of a majority must stand as the judgment of the whole. This is a law imposed on us by the absolute necessity of the case; and if we do not act upon it, there is no possibility of maintaining any government but despotism. (Webster VI:219)[38]

Webster's characterization of the consequences of nullification as minority rule or despotism is lent plausibility by an earlier reference to one of the founders. After showing how nullification would return the nation to its abhorrent pre-Constitution conditions, he quotes Oliver Ellsworth as deploring the despotic, instead of the more apparent anarchic, tendencies of those times.

> In republics, it is a fundamental principle, that the majority govern, and that the minority comply with the general voice. How contrary, then, to republican principles, how humiliating, is our present situation! A single State can rise up, and put a veto upon the most important public measures . . . So far is this from being consistent with republican principles, that it is, in effect, the worst species of monarchy. (Webster VI:208)[39]

Both Ellsworth and Webster, however, presume the propriety of national majorities. They must admit that a state veto does not necessarily violate majority rule on the state level. In fact, Calhoun's open hope is to (re-)institute a federal system which is more sensitive to state majorities. He hardly despises the people of South Carolina and their elected representatives. Webster, again, conveniently slides over the difference between his opponent's positions on the state and federal levels. He calls attention to that difference by referring to the South Carolina unionists as an oppressed minority. Yet, he professes to only see that as an inconsistency on Calhoun's part, as a blemish on his vaunted solicitude toward minorities (see Webster VI:221–22). The deeper consistency to Calhoun's position is not that he peculiarly distrusts majority rule but that he peculiarly distrusts majority rule in a large, diverse nation.[40]

Calhoun is enough of a nationalist to accept qualified national majorities, such as occur in the amending process. Even Webster, notwithstanding Calhoun's accusations to the contrary, does not favor absolute national majorities (see Calhoun II:306).[41] In any case, Calhoun does not reject majority rule *per se,* nor does he reject its underlying assumptions. He "merely" considers those assumptions more valid on the state level. In this respect, too, his differences with Webster are the differences between a federal and a national republican, not a nonrepublican and a republican. Their reverse-image descriptions of the current political situation in South Carolina provide crucial evidence of the true character of their differences. Those descriptions attest to a shared republican idiom with conflicting empirical referents.

Their debate, though, hinges on the nature of the union as much as on the nature of the nullification movement. Both statesmen agree that the union exhibits at least some of the features of a national republic. They disagree over how many and, therefore, over whether federal or national republicanism is more problematic under American conditions. A fundamental component of Calhoun's federal republicanism is his legitimation of the American states by the liberal ends of liberty and happiness. In justifying the union by those same ends, Webster, similarly, reinforces his national republicanism. Once more, there are strong parallels between Webster and Publius, but also substantial differences.

Liberty and Union

Webster's whole argument does stand on his defense of the union; on the value he can assign, and can convince others to assign, to the union. The most famous of all his turns of phrase is: "Liberty and Union, now and for ever, one and inseparable!" (Webster VI:75). The peroration of his "Second Reply to Hayne" etched this claim on the collective consciousness of a generation of American statesmen. Its meaning, however, is far from obvious. Webster is still struggling with its meaning in his nullification speech.[42]

The meaning of the claim seems obvious. Liberty and union are one because liberty is the defining characteristic of the union. They are inseparable because an exemplary amount of liberty would not exist in the American territories if they were disunited. This exemplariness is what, in turn, makes liberty the defining characteristic of the union.

Webster's meaning, though, is not Publius's. His claim is much more concrete; much less the conclusion of a deductive means-ends argument. For him, it is a particular form of liberty that is one and inseparable with the union. He sets up a determinate relationship between liberty and union with the Constitution as the middle term that secures both. He calls the particular form of liberty thus wedded to the Constitution and the union "constitutional" or "American" liberty. He rebukes Calhoun for discussing liberty in more abstract terms.

> The honorable gentleman has declared, that on the decision of the question now in debate may depend the cause of liberty itself. I am of the same opinion; but then, Sir, the liberty which I think is staked on the contest is not political liberty, in any general and undefined character, but our own well-understood and long-enjoyed American liberty. (Webster VI:183)

This rebuke merges into his methodological critique of Calhoun's Senate resolutions for being too philosophical and for not being sufficiently sensitive to the uniqueness of American political experience. Webster articulates his own cause in such a way as to avoid those "mistakes."

> Yet, Sir, it is our own liberty, guarded by constitutions and secured by union, it is that liberty which is our paternal inheritance, it is our established, dear-bought, peculiar American liberty, to which I am devoted, and the cause of which I now mean, to the utmost of my power to maintain and defend. (Webster VI:183)

The concreteness of Webster's claim directly follows from his organic view of the union. According to him, liberty and union are not two abstract terms in a political equation. They are *this* liberty and *this* union which have grown inseparably together and which would, in his worst nightmares, be irretrievably destroyed together. His claim, however, does not put any particular weight on the amount of liberty which is peculiarly American. Unlike Publius, Webster does not deny that disunion might result in a greater liberty in the states—which is the suppressed premise of Calhoun's concurrent-majority theory. He only denies that it would be the same liberty. His claim is tautologically true. Yet, it leaves him open to criticisms that Americans could enjoy more liberty under a variety of different political arrangements.[43]

Webster, nevertheless, does believe that American liberty is a quantitatively great form of liberty, greater than that which exists, or has existed, in any other nation. This superiority is implied by America's world-historic mission, something else he accuses Calhoun of threatening.

> Mr. President, if the friends of nullification should be able to propagate their opinions, and give them practical effect, they would, in my judgment, prove themselves the most skillful "architects of ruin," the most effectual extinguishers of high-raised expectation, the greatest blasters of human hopes, that any age has produced. They would stand up to proclaim, that the last great experiment of representative government had failed. They would send forth sounds, at the hearing of which the doctrine of the divine right of kings would feel, even in its grave, a returning sensation of vitality and resuscitation. Millions of eyes, of those who now feed

> their inherent love of liberty on the success of the American example, would turn away from beholding our dismemberment, and find no place on earth whereon to rest their gratified sight. (Webster VI:236–37)[44]

Consequently, Webster does not think that American liberty can be improved by disunion. Even if, against all probability, disunion were to lead to a greater liberty in the states, it would be a qualitatively lesser liberty. It would possess less exemplary value. It would not be encased in the unprecedented experiment of a representative government extending over a large empire. Webster contends that America holds the world's attention both because of "the character of her institutions" and because of "her power, her rapid growth, and her future destiny" (Webster VI:227). Disunion (or any other systemic change?) would destroy the force of her example; in effect, her liberty. Webster predicts that "[a]midst the incantations and orgies of nullification, secession, disunion, and revolution, would be celebrated the funeral rites of constitutional and republican liberty" (Webster VI:237).

Calhoun would not disagree that disunion destroys the force of the example. But he, on the one hand, denies that nullification means disunion. On the other hand, he understands the nature of the American mission and of American institutions in very different terms. For him, the mission is essentially defined by the federal features of the union. It, therefore, can be subverted by either consolidation or disunion.[45] Webster, like Publius, only seems to admit the latter possibility. They understand the mission as being essentially nationalistic, both in institutional and ideological terms. Yet, what for Publius was an inherently progressive force becomes for Webster a conservative one. The mission means preserving the changes Publius advocated rather than advocating further changes (see Webster VI:226, 237).[46] The fact that the union now has a (longer) history seems to structure Webster's entire political argument.

Webster's treatment of "happiness and union" betrays a similar, conservative, or, perhaps better, historicist, bias. He claims that the American people currently enjoy a high level of public and private felicity. He insists that the union, as perfected by the Constitution, is one of the causes. He, finally, expresses incredulity that the nullifiers would jeopardize the union and their own political happiness over such a relatively minor issue as the federal tariff rates.

> And now, Mr. President, what is the reason for passing laws like these [South Carolina's nullification ordinances]? What are the oppressions experienced under the Union, calling for measures which thus threaten to sever and destroy it? What invasions of public liberty, what ruin to private happiness, what long list of rights violated, or wrongs unredressed, is to

> justify to the country, to posterity, and to the world, this assault upon the free Constitution of the United States, this great and glorious work of our fathers? At this very moment, Sir, the whole land smiles in peace, and rejoices in plenty. A general and a high prosperity pervades the country and, judging by the common standard, by increase of population and wealth, or judging by the opinions of that portion of her people not embarked in these dangerous and desperate measures [the unionists], this prosperity overspreads South Carolina herself. (Webster VI:226)

Embedded in this passage is a long list of ends—public liberty, private happiness, (the absence of) rights violated and wrongs unredressed, peace, plenty, a general and high prosperity, and an increase of population and wealth—which define the political happiness of the American people.[47] There, however, is no argument for how the union secures those ends, such as we found in *The Federalist Papers*. There is only the bare assertion that they are secure in union, including, importantly, in South Carolina. We also wonder what happens when they conflict. Overreaching Publius (and far overreaching the Federal Farmer and Calhoun!), Webster exalts a beneficent union not so much to deny the possibilities of conflict as to trivialize them.

Webster appears to place the historicity of the union beyond the nature of the union, even beyond the value of the union.[48] The qualities he uses to describe the nature and value of the union are identifiably republican ones. Even if he was not aware of those qualities as republican ones, he must have been aware of how roughly they fit their object. He, and his nationalist successors, struggled to maintain contact between the qualities and the object. Their efforts provided the central dynamic of American political thought up to the Civil War, a dynamic which, almost imperceptibly, spurred a continuing shift from republican to pluralist ideas. In the meantime, though, they could always magnify the object. As one of Webster's last pleas in his nullification speech admonishes his audience, "We cannot, we must not, we dare not, omit to do that which, in our judgment, the safety of the union requires" (Webster VI:237).

PART THREE

THE LINCOLN-DOUGLAS DEBATES

The natural tendency is to place Lincoln, Webster, and Publius in one intellectual tradition (or subtradition) and Douglas, Calhoun, and the Federal Farmer in another. This tendency is so natural because it makes so much sense of the "data." The major dimensions of the Lincoln-Douglas debates, however, were not the major dimensions of the ratification and nullification debates. Issues of slavery and democracy had at least partially superseded issues of federalism and nationalism. The underlying arguments were also different in character. They were now more pluralistic and less republican. While the issues and arguments would look somewhat different if we were to analyze an intersectional, instead of an intrasectional, debate over slavery, they would not look that different. Furthermore, by the late 1850s the key question was not whether the North and South had become increasingly alienated from each other but whether the more moderate groups within each section could maintain their political ascendancy over the more radical groups. In the North, the most politically significant debates were between the Douglas Democrats and the Lincoln Republicans.

7

Stephen A. Douglas, Slavery, and Pluralist Democracy

Stephen A. Douglas's doctrine of popular sovereignty seems locked into the federal-republican tradition. And so it is, on the surface. Douglas marshals the same small-republic argument against Lincoln as Calhoun and the Federal Farmer marshaled against Webster and Publius. According to Douglas, the union is so diverse that the only thing which holds it together is the initial agreement to allow different states and territories to make their own decisions on those matters which divide them from each other. His popular-sovereignty doctrine appears to be the *reductio ad absurdum* of states' rights.

Douglas's definition of the union, however, seems even more pluralistic than Calhoun's or the Federal Farmer's. This contrast obviously reflects the fact that the union *is* more pluralistic than it was in 1833 or in 1788. On the slavery issue, for instance, a "necessary evil" consensus has given way to a range of positions from immediate abolition to slavery expansion. Douglas, though, also appears to be responding to the extraordinary economic and ethnic diversity of white America at midcentury.

Ironically, the same realism which undergirds Douglas's federal republicanism undermines it. Bleeding Kansas is not nullifying South Carolina. Douglas only emphasizes the negative side of the small-republic argument. His states and territories appear almost as pluralistic as his union does.[1]

As always, it was not quite that simple. In order for Douglas to have offered his popular-sovereignty doctrine as a viable solution to the slavery issue, he must have made two assumptions. First, he must have assumed that the American states and territories were much more homogeneous on the issue than the union, as a whole, was. But would not the states and territories have to be continually subdivided to assure pro- or antislavery consensuses? In that case, Douglas's doctrine would indeed have been the *reductio ad absurdum* of states' rights.[2] He, then, must have made a second assumption. He must have believed that the dynamics of popular sovereignty would make the states and territories more homogeneous over time through internal migration. This assumption, though, implied that all American citizens were equally mobile. The Southern slaveholders, of course, denied the validity of that implication. They claimed that an early decision in favor of slavery in the territories was critical to the efficacy of their right of movement. Douglas, nonetheless, left the locus and timing of popular sovereignty ambiguous. This ambiguity was politically expedient in enabling him to appeal to moderate Democrats in both the North and South.[3] Yet, he also left those matters ambiguous because the viability of his solution to the slavery issue did not really depend on the traditional federal-republican assumption of intrastate homogeneity or, correspondingly, on states' rights. It, rather, depended on a nontraditional, pluralistic theory of democracy for which the locus and timing of popular sovereignty were less important than the duties the doctrine prescribed for American citizens, whether acting within or across state boundaries.

According to Douglas, the primary means of effecting any solution to the slavery issue is for citizens to profess mutual indifference to each others' preferences on the issue. This duty is more positively stated as the mutual respect of citizens for each others' preferences and the mutual self-restraint of citizens toward each others' preferences or, in short, interpersonal comity. For Douglas, interpersonal comity is part of a broader theory of democracy. The basic intuition behind this theory is that it is undemocratic for citizens to seek to impose their preferences on other, equal citizens either through force, their legal or personal authority, or even persuasion. The radical sense in which Douglas understands self-government distinguishes his theory of democracy as a pluralistic one. Douglas's democracy is a matter of constructing majorities through the addition and subtraction of autonomous preferences, not through a public dialogue between those who hold conflicting preferences on a given issue. It is essentially a democracy of silence, including the (self-)muzzling of statesmen who, in particular, might be able to impose their preferences on other citizens. It is a democracy on the model of a free-market economy, in which statesmen have largely negative duties to protect the integrity of the process and *not* to try to deflect it in any particular direction. It, finally, is a democracy which is antithetical to traditional republican views of citizenship, statesmanship, and popular government.

Where someone lives still matters in Douglas's pluralistic theory of democracy. Citizens with similar preferences want to live together so that they are more likely to be in the majority on any given issue. Douglas's theory of democracy has its own federalist corollary. His commitment to federalism and states' rights, however, is largely loosened from traditional republican concerns. States' rights merely provides a convenient decision-rule in a pluralist democracy. Pluralism lowers the stakes of states' rights and a highly mobile, rapidly expanding society dissipates most problems of preference coordination. States' rights becomes understood as interstate comity—a correlative duty to interpersonal comity—and it loses its specific geographic focus. As a result, Douglas could celebrate his nationalism in a much less ambivalent way than either Calhoun or the Federal Farmer could. His theory of democracy seems fundamental to his stances on federalism, nationalism, and slavery. The converse view has, of course, been promulgated by the prevailing interpretations of his political thought.

This chapter constitutes an extended argument against the view that Douglas's commitment to states' rights was fundamental to his theory of democracy. Here, I will briefly address two other opposing views: (1) that his theory of democracy was primarily a politically advantageous response to his generation's escalating crisis over the slavery issue; (2) that it was, in a less self-serving fashion, a response to his nationalistic desire to hold the union together in the face of that crisis.[4]

Douglas clearly thought popular sovereignty was a doctrine that could attract considerable support in both the North and the South, recementing the Democratic Party and easing his way into the White House. No one would deny that he was strongly committed to his party and to his own political advancement within it. Similarly, no one would deny that he was strongly committed to the nation and its manifest destiny. He also thought of popular sovereignty as the only doctrine which could preserve the union and allow for its continued expansion.

The problem with these two explanations of Douglas's popular-sovereignty doctrine is that they do not distinguish him from his equally ambitious, equally partisan, and equally nationalistic rivals, such as Abraham Lincoln. The thesis that his doctrine was rooted in a broader, and contested, theory of democracy does that work. The three competing (though not mutually exclusive) explanations also answer somewhat different questions. While the "political ambition" and "save the union" explanations focus on the question of what popular-sovereignty doctrine did for Douglas, the "pluralist democracy" explanation focuses on the question of where it came from, not in the sense of its personal pedigree but of its intellectual foundations.[5]

Douglas's theory of democracy went quite deep. It helped determine how he would pursue his political ambition, how he would react to the slavery issue, and how he would seek to save the union. This is not to say that the various positions Douglas took during the course of his debates with Lincoln were derived deductively from a particular theory of democracy. They were, more, based on an increasingly popular, intuitive understanding of democracy. Yet, that understanding did anchor those positions.

Popular Sovereignty

Lincoln kicked off his campaign to unseat Douglas in Springfield on 16 June 1858 with a speech accepting his nomination as the Republican Party's senatorial candidate. This speech has come to be known as the "House Divided" speech in reference to its opening claim that the union cannot long endure as a house divided against itself, half free and half slave. We will explore what exactly Lincoln meant by this claim in the next chapter. For now, the important thing is that it became the focal point of the ensuing campaign, including Douglas's initial campaign speech three weeks later in Chicago and the seven formal debates between the two Illinois statesmen.[6]

Douglas contends that Lincoln's "House Divided" doctrine is a revolutionary one. It means disunion and civil war. The union, to the contrary, can (and will?) remain a house divided forever, just as the founding fathers made it and just as all the major political parties in America had labored to preserve it until the "black" Republicans set out to abolitionize the "old-line" Whigs and Democrats. Underlying this counterargument is the much broader, and older, argument that the union is too diverse to be a house united on many (most?; almost all?; all?) issues. According to Douglas, the founders shared this understanding as they went about the process of creating a federal system of government which would secure not only the union but the liberty, prosperity, and destiny of the American people. He repudiates Lincoln's "House Divided" doctrine for mandating national uniformity, thus threatening everything Americans hold most dear.

In a significant sense, Douglas does simply plug the slavery issue into the traditional federal-republican apparatus. By 1858, it is *the* example of interstate diversity.[7] The Lincoln-Douglas debates seem to follow the well-worn paths marked out by previous states-rights debates, with the increased stakes of a nation splitting apart over the slavery issue. Douglas attempted to lower the stakes by making it a local issue while accusing Lincoln of imprudently raising the stakes by making it a national issue.

In his Chicago address, Douglas sets out his basic argument against the "House Divided" doctrine.

> I assert that it is neither desirable nor possible that there should be uniformity in the local institutions and domestic regulations of the different

> States of this Union . . . The Fathers of the Revolution, and the sages who made the Constitution, well understood that the laws and domestic institutions which would suit the granite hills of New Hampshire would be totally unfit for the rice plantations of South Carolina; they well understood that the laws which would suit the agricultural districts of Pennsylvania and New York would be totally unfit for the large mine regions of the Pacific, or the lumber regions of Maine. (Johannsen, p. 29 [Douglas])

He then draws, or, rather, has the founders draw, the constitutional implications of this argument.

> They well understood that the great varieties of soil, of production and of interests, in a Republic as large as this, required different local and domestic regulations in each locality, adapted to the wants and interests of each separate State, and for that reason it was provided in the Federal Constitution that the thirteen original States should remain sovereign and supreme within their own limits in regard to all that was local, and internal, and domestic, while the Federal Government should have certain specified powers which were general and national, and could be exercised only by federal authority. (Johannsen, pp. 29–30 [Douglas])[8]

Douglas is obviously determined to show that slavery is a local question under the founders' federal system. He even suggests he and Lincoln disagree only over where the slavery issue belongs under that system (see Johannsen, pp. 326, 329 [Douglas]). Throughout the course of their debates, Douglas, nevertheless, assumes their differences are more general. He presents Lincoln as an advocate of consolidation and national uniformity and himself as the defender of federalism and local diversity. He considers those differences "direct, unequivocal, and irreconcilable" (Johannsen, p. 34 [Douglas]). They are also extremely significant insofar as he believes consolidation endangers, and federalism protects, the fundamental ends of the American people.

The prior, but not necessarily primary, political end is the union itself. Douglas indicts Lincoln's "House Divided" doctrine for fomenting disunion. "What does Mr. Lincoln propose? He says that the Union cannot exist divided into free and slave States. If it cannot endure thus divided, then he must strive to make them all free or all slave, which will inevitably bring about a dissolution of the Union" (Johannsen, p. 73 [Douglas]). Worse, it is an incitement to civil war. "In other words, Mr. Lincoln advocates boldly and clearly a war of sections, a war of the North against the South, of the free States against the slave States—a war of extermination—to be continued relentlessly until the one or the other shall be subdued, and all the States shall either become free or become slave" (Johannsen, p. 29 [Douglas]). Alternatively, Douglas's (and the founders') popular-sovereignty doctrine will maintain peace and harmony between the states. "When that principle is recognized, you will have peace and harmony and fraternal feelings between all the States of this Union; until you do recognize that doctrine, there will be sectional warfare agitating and distracting the country" (Johannsen, p. 73 [Douglas]). In fact, the union can endure indefinitely as a house divided, just as it has for the past seventy years. "I then said, I have often repeated, and now again assert, that in my opinion our Government can endure forever divided into free and slave States as our fathers made it,—each State having the right to prohibit, abolish or sustain slavery, just as it pleases" (Johannsen, p. 288 [Douglas]).

Needless to say, Douglas understands the union as a federal union. He puts himself in the same, somewhat paradoxical, position as Calhoun and the Federal Farmer in arguing that the preservation of the union depends on the degree to which it is *not* politically centralized and in appearing to value the union only because he is emotionally attached to it. Logically, little seems to separate him from the Northern abolitionists and Southern fire-eaters who are clamoring for disunion in order to maintain sectional purity.

Most important, American liberty does not seem to require union. Again like Calhoun and the Federal Farmer, Douglas ties liberty directly to the states, not the union. He believes American citizens will collectively enjoy more liberty within a federal than within a consolidated union because each citizen will be more likely to find himself in a local than in a national majority. Federalism allows more citizens to satisfy their preferences. It, therefore, is an integral means to their right of self-government in both the public and private senses of the term. These claims are all implicit in the following passage from Douglas's Chicago speech.

> Uniformity in local and domestic affairs would be destructive of State rights, of State sovereignty, of personal liberty and personal freedom. Uniformity is the parent of despotism the world over . . . Wherever the doctrine of uniformity is proclaimed, that all the States must be free or all slave, that all labor must be white or all black, that all the citizens of the different States must have the same privileges or be governed by the same regulations, you have destroyed the greatest safeguard which our institutions have thrown around the rights of the citizen. (Johannsen, p. 30 [Douglas])

It is true that Douglas calculates the value of the union. In certain respects, it, too, is an integral means to the political happiness of the American people. The diversity of the union adds to the national welfare; disunited, Americans would not be nearly as prosperous (see Johannsen, pp. 48, 197 [Douglas]). Similarly, the nation's world-historic mission for freedom depends on the continued existence of the union, as does the fulfillment of its manifest destiny (see Johannsen, p. 105 [Douglas]). These, by now, are familiar arguments which unite federal and national republicans.

Even in these respects, though, the linchpin of Douglas's arguments is the existence of the union as a federal union. Both the national welfare and the American mission require federalism. Consolidation, conversely, spawns despotism and blocks expansion (see Johannsen, pp. 30–31, 129–30 [Douglas]).[9] Douglas's conclusion is a traditional federal-republican one. Strongly federal principles sustain the union and those principles, not the intrinsic qualities of the union, are responsible for the unparalleled blessings of the American people. Douglas exalts the consequences of the nation's purported commitment to those principles—to his own popular-sovereignty principles—in the opening speech of the debates.

> Under that principle, we have grown from a nation of three or four millions to a nation of about thirty millions of people; we have crossed the Allegheny mountains and filled up the whole North-west, turning the prairie into a garden, and building up churches and schools, thus spreading civilization and Christianity where before there was nothing but savage barbarism. Under that principle we have become, from a feeble nation, the most powerful on the face of the earth, and if we only adhere to that principle, we can go forward increasing in territory, in power, in strength and in glory until the Republic of America shall be the North Star that shall guide the friends of freedom throughout the civilized world. (Johannsen, p. 48 [Douglas])[10]

Douglas's reading of American history also places him within a federal-republican tradition. He insists that popular-sovereignty principles have indeed governed American political practice up until that time. He, thus, claims that the founders wrote those principles into the Constitution, apparently overlooking the cases where they prejudiced national over local majorities as well as the fact that they did not clearly make slavery either a national or a local question. The contrary position, as defended by Lincoln, is that even though the Constitution does not permit the federal government to regulate slavery in the states, it does permit the federal government to regulate slavery in the territories (see Johannsen, pp. 131–32, 254–55, 315 [Lincoln]).[11] Douglas offered a detailed rebuttal of that position in an 1859 *Harper's Magazine* article. In his debates with Lincoln, he simply assumes the conclusion of that article, that states and territories possess (almost) equal rights under the Constitution.[12]

A corollary to Douglas's position is that the founders made the house divided (see Johannsen, pp. 44, 104, 125, 197, 218, 288, 326 [Douglas]). According to Douglas, the union then had twelve slave states and only one free state. If the founders had acted on Lincoln's "House Divided" doctrine, it would have meant "slavery national," not "freedom national" (see Johannsen, pp. 44–45, 218, 289 [Douglas]).[13] For Lincoln and the Republican Party to now try to impose "freedom national" on the union—just because the free states have finally achieved a majority—would not, as alleged, resurrect the policies of the founders. It would seriously violate them (see Johannsen, pp. 218, 289–90 [Douglas]).[14] Douglas attests to his own filiopiety by promising "to sustain the Constitution of my country as our fathers made it" (Johannsen, p. 32 [Douglas]).

Jumping forward in history, Douglas contends that during the second-party system Democrats and Whigs equally embraced the doctrine of popular sovereignty. This consensus produced the Compromise of 1850, when members of both parties rallied around Clay, Webster, and Cass to quell sectional agitation (see Johannsen, pp. 38, 98, 116–17, 186–87, 216, 296, 325 [Douglas]).[15] Not surprisingly, Douglas views his own Kansas-Nebraska Act of 1854 as another result of that consensus (see Johannsen, pp. 26, 38–39, 206, 216, 296 [Douglas]).[16] *Dred Scott* is more problematic and his efforts to adjust popular-sovereignty doctrine to Taney's decision are justly famous, or infamous (see Johannsen, pp. 88–89, 160–61, 217–18, 269–70, 328–29 [Douglas]).[17] In the immediate context of 1858, Douglas positions himself between the Buchanan Democrats, whom he considers disloyal to their own party principles in attempting to force the proslavery Lecompton constitution on an antislavery majority in Kansas, and the Republicans, whom he considers a purely sectional party. Although the Republicans helped Douglas oppose the administration's Lecompton fraud, he suggests they did so primarily on antislavery grounds and only incidentally on popular-sovereignty grounds (see Johannsen, pp. 24, 130, 292 [Douglas]). He believes the positions of both the Buchananites and Republicans are not only immoderate but heretical.[18]

For Douglas, then, history, constitutional law, and democratic theory converge on popular sovereignty. His positions on all three dimensions closely parallel those of earlier federal republicans, such as Calhoun and the Federal Farmer. The subtext, though, is very different.

Douglas does not present federalism as a historical and constitutional reflex to the existence of small, homogeneous states within a large, diverse union. While he uses national diversity to attack the "House Divided" doctrine, that strategy does not lead him to assert social homogeneity on the state level. It, instead, leads him to celebrate diversity, and the protection of diversity, as necessary to free government. It leads him to develop a pluralistic theory of democracy which does not essentially distinguish between the states and the union.

Within this theory, the argument for federalism proceeds on quite different grounds. Douglas's popular-sovereignty doctrine realizes the advantages of grouping like-minded citizens and drawing political boundaries around such opinion blocs. Those boundaries, however, will never define zones of unanimity. Unlike his predecessors, Douglas sees the problems of democratic decision making within the states as basically the same as the problems of democratic decision making within the union. Unlike them, his solution does not depend on political geography. He is much less interested in developing a constitutional theory which underwrites a strongly federal system of government than he is in developing a democratic theory which can be applied on a large, national scale.

Not wholly, but for the most part, Douglas transcends the tensions between federalism and nationalism within the American republican tradition by transcending that tradition. He does not portray either the union or the states in traditional republican terms. His pluralism enables him to advocate federalism as an optimal decision-rule in a large, diverse society and yet to embrace nationalism by circumventing the traditional problems of scale.

The Democratic Intuition

The basic message behind Douglas's doctrine of popular sovereignty is that states (and territories) should not interfere with each others' institutions. Again and again, he affirms "the right of each state to do as it pleases, without meddling with its neighbors" (Johannsen, p. 129 [Douglas]).[19] This message is easily misinterpreted for it makes popular sovereignty appear to be a theory exclusively about the Constitution and federalism, and not also about democracy. In this section of the chapter, I will argue that Douglas's injunction against states interfering with each others' institutions is derived from a much broader intuition which understands the principle of noninterference quite literally and which, therefore, also enjoins citizens from interfering with each others' affairs independent of state boundaries.

Lincoln, after all, would not disagree with the principle of noninterference in a legal sense or in the sense of forbidding one state from taking overt action against the institutions of another state. Such forbearance certainly is also part of his constitutional theory. Douglas, however, understands the principle of noninterference in a much more radical sense so that it means the citizens of one state can barely discuss the institutions of other states. Lincoln naturally resists this extension of the principle.[20]

Thus, Douglas declares not just that states should not meddle with their neighbors, but that "it is none of our business in Illinois whether Kansas is a free State or a slave State," that "when we settled it [the slavery question] for ourselves, we have done our whole duty," and that no other state has the right "to complain of our policy in that respect, or to interfere with it, or to attempt to change it" (Johannsen, pp. 292, 47, 33 [Douglas]). He charges that much of Lincoln's rhetoric during the course of their debates amounts to complaining about, interfering with, or attempting to change the peculiar institutions of the Southern states (see Johannsen, pp. 47–48, 128–29, 266–67, 276, 329 [Douglas]).

The end Douglas has in mind in enjoining citizens from interfering with—in this fairly radical sense of interfering with—the institutions of other states is peace and harmony between the states (see Johannsen, pp. 73, 161, 219, 275, 300 [Douglas]). His views on interstate comity clearly are the views of someone who is firmly committed to states' rights. He seems intent on establishing an important distinction between how American citizens should act within and across state boundaries. Nevertheless, his views on interstate comity suggest equally radical views on interpersonal comity. If citizens should not interfere with the institutions of other states in order to maintain peace and harmony between the states, they also should not interfere with the institutions of their own states or, more to the point, with the affairs of other citizens in their own or other states in order to maintain peace and harmony throughout the union.

Douglas's views on interstate comity imply that the best way to handle political controversy, wherever it occurs, is silence, or only speech of a certain, deflationary sort which allows the controversy to be quieted through noncontroversial processes of aggregating existing preferences on the issue(s) involved in the controversy. The reason he, then, distinguishes between how citizens should act within and across state boundaries is that he assumes there is less controversy within than between states. However, he also recognizes the limits of that assumption. In Illinois, he is his own witness to the fact that the slavery issue has generated plenty of controversy.[21] He ultimately suggests that the citizens of Illinois who hold conflicting opinions on the issue should not act any differently toward each other than they do toward the citizens of other states who hold conflicting opinions. In sum, Douglas's democratic intuition conflates interstate and interpersonal comity.

Douglas's views on statesmanship, and, in particular, his views on how such Illinois statesmen as Lincoln and himself should act, bring us closer to the core of his democratic intuition. Once more, he starts from a states-rights premise. He asserts that "no political creed is sound which cannot be proclaimed fearlessly in every State of this Union" (Johannsen, p. 211 [Douglas]).[22] He, thus, sets up a criterion of national statesmanship which decrees that the less controversial a doctrine is across state lines, the better it is. Judged by this criterion, his popular-sovereignty doctrine is clearly superior to Lincoln's "House Divided" doctrine. Presumably, no political doctrine can be less controversial across state lines than the doctrine that "each State of this Union has a right to do as it pleases on the subject of slavery." It surely will be less controversial than a doctrine that argues "the question whether slavery is right or wrong" (Johannsen, p. 266 [Douglas]). According to Douglas, states-rights doctrines will always trump national-uniformity doctrines *on nationalist grounds.*

At this point, it will be useful to briefly examine two of Lincoln's responses to Douglas's views on statesmanship because they serve to further expose the core of his democratic intuition. First, Lincoln notes that Douglas, in refusing to argue the right or wrong of slavery, is refusing to argue the question which most concerns most Americans (see Johannsen, p. 315 [Lincoln]). Second, Lincoln observes that Douglas cannot proclaim his popular-sovereignty doctrine everywhere; he cannot proclaim it in Tsarist Russia (see Johannsen, p. 222 [Lincoln]). Both responses stress the controversial nature of even Douglas's "neutral" rhetoric. When people really care about an issue, neutrality is itself controversial. Moreover, what is neutral differs in different societies. Douglas retorts that he only said he could proclaim his popular-sovereignty doctrine everywhere in the United States (see Johannsen, p. 238 [Douglas]). But is even that true? Can he proclaim his doctrine to an abolitionist audience in Syracuse, New York, or to a proslavery audience in Charleston, South Carolina? And if he cannot, will he become silent everywhere in the United States or, alternatively, change his doctrine? Lincoln's lesson is twofold: every doctrine is bound to be controversial somewhere and statesmen do not necessarily avoid controversy. He is not really inviting Douglas to go to Russia to preach the evils of Tsarist despotism; he is inviting him to preach the evils of racial slavery throughout the United States.

If Lincoln discloses the limits of Douglas's statesmanship, Douglas returns the favor in full. Douglas insists not only that Lincoln cannot proclaim his "House Divided" doctrine throughout the United States but that he cannot even consistently do so throughout his own state. Although his principles are "jet-black" in northern Illinois, by the time he has been trotted down to "Little Egypt" they have turned "almost white" (Johannsen, p. 195 [Douglas]).[23] Lincoln naturally denies the charge and there seems to be little substance to it (see Johannsen, pp. 198, 220–22, 247–49 [Lincoln]).[24] Yet, it does allow Douglas to reassert his claim that Lincoln's statesmanship has a very geographically limited audience and that he himself does not need to trim his message because he has a statewide, indeed a national, audience.

States' rights, however, no longer seems to constitute an effective constraint on the rhetoric of American statesmen for Douglas except insofar as it is the only message which he, rightly or wrongly, thinks can gain a national audience. In the end, he appears more concerned with avoiding political controversy than with respecting states' rights and more concerned with protecting a national party coalition than with cultivating state or local majorities. He preaches the blessings of popular sovereignty wherever he goes in America because he firmly believes that message minimizes controversy and stretches the boundaries of his audience. His views on statesmanship are not primarily about a federal union but about a national democracy.

Furthermore, Douglas's views on statesmanship assume a particular understanding of democracy. It is that understanding which induces him to present the relationship between statesman and audience as a reflexive and static one, as one in which the preferences of the audience are taken as given and the statesman's role is to articulate the majority preferences. Obviously, it can never be that simple and we should resist the temptation to caricature Douglas's views. They do not preclude statesmen from baring latent preferences, from trying to construct majorities out of conflicting preferences, or from forging policies which creatively translate majority preferences into law. Douglas himself cannot consistently act as a mere cipher for majority preferences. He proudly relates the story of how he calmed a Chicago "mob" which had gathered in angry protest against the passage of his Kansas-Nebraska bill (see Johannsen, pp. 156–57 [Douglas]). It, however, is significant that he calls it a mob (instead of a group of citizens with opposing ideas about how to best organize the new territories) and that he says he was able to persuade it of the merits of the bill and of the bill's underlying popular-sovereignty principles (proving that a latent consensus on popular sovereignty existed which he merely brought to the surface). Douglas also modestly acknowledges that he has done more than anyone to make popular sovereignty the law of the land (see Johannsen, p. 26 [Douglas]). But, again, the initial impression of active statesmanship does not survive a second look. In convincing Americans to embrace popular-sovereignty principles, Douglas is really convincing them to honor the legitimacy and autonomy of their own preferences and to make their own decisions on the controversial issues of the day without any substantive guidance from himself or anyone else. He, therefore, envisions relatively narrow limits on what actions statesmen may take in a democracy to aggregate individual preferences.

Douglas's continuing attack on Lincoln's statesmanship reveals those limits. Among other things, he accuses Lincoln of promising what he cannot deliver, at least not in a democracy. He, thus, asks Lincoln what he proposes to do to abolish slavery in the South and to overturn the *Dred Scott* decision if he is elected senator (see Johannsen, pp. 243, 265–67, 327 [Douglas]). He characterizes the alternatives facing Lincoln, or any statesman in a similar situation, as either promoting immediate, revolutionary action against institutions they consider evil or consigning themselves to silence and the acceptance of existing evils as practical necessities.[25] He seems to overlook the obvious possibility that by doing nothing but talking about the evils of existing institutions statesmen might, over time, construct majorities which will act to abolish those institutions through democratic means. Douglas's narrowing of his opponent's options is good politics. It, however, is also entailed by his understanding of democracy.

Here, it is the dynamic dimension of democratic politics which seems to escape Douglas. Despite his professed indifference to the end of slavery, he did look forward to the day when the preferences of most Southerners had changed and they themselves abolished their evil institution. Yet, if he can rebuke Lincoln for not stating exactly how slavery will be abolished, all the more can Lincoln rebuke him for not even publicly stating his desire that it be abolished.[26] At a minimum, they have very different expectations about how the institution of slavery will come to an end in America. Lincoln believes slavery will be abolished as the result of an extended, democratic dialogue over the evils of the institution, a dialogue in which morally concerned statesmen will play a central role. Douglas, in contrast, believes it will be abolished as the result of a more natural process of opinion change. He, for instance, claims that the end of slavery in Illinois was *not* the result of an extended, democratic dialogue over the evils of the institution but rather of an apparently spontaneous process whereby the citizens of Illinois came to realize that it no longer served their interests (see Johannsen, p. 299 [Douglas]). Presumably, the same process will some day end slavery in the South; what he, or Lincoln, might say about the evils of the institution will not contribute to that process (and might well deflect it). My primary interest, though, is not in their different scenarios for the end of slavery, nor in their similar preferences for its ultimate demise. I am more interested in their different understandings of what is democratically permissible in trying to bring about its demise.

What Douglas finds so undemocratic about Lincoln's politics is the image, if not the reality, that he is attempting to impose his antislavery preferences on other, equal citizens. This image violates Douglas's fundamental intuition that democracy means interpersonal comity. Because those in positions of power possess the greatest potential to violate interpersonal comity, they possess the strongest duties to practice it. As Douglas understands it, interpersonal comity prescribes a set of duties for both statesmen and citizens. It enjoins them to virtual silence about their own policy preferences, feigned (or, perhaps worse, unfeigned) indifference to policy decisions, and absolute devotion to maintaining the integrity of a democratic process that simply aggregates individual policy preferences.

The fact that policy decisions must (sometimes) be made seems to destroy the illusion of Douglas's noncoercive democratic process. Do not public policies represent the imposition of majority preferences on the minority? To Douglas, however, that "imposition" is ideally an impersonal one and, therefore, not really an imposition.[27] As long as the process simply adds and subtracts individual preferences, members of the minority have merely been outvoted by other, equal citizens. The legitimacy and autonomy of their preferences have not been impugned. Their preferences have been accorded equal consideration right through the time a policy decision was made. Douglas's counterexample is, again, the efforts of the Republican Party, with its anticipated sectional majority, to impose "freedom national" on a recalcitrant Southern minority. Beyond violating states' rights, that policy is undemocratic to the extent that it denies the legitimacy and autonomy of Southern proslavery preferences.

Douglas sets conditions on democratic statesmanship which no statesman could, or would wish to, religiously follow in practice. Douglas certainly did not, though he struggled mightily to do so. Those conditions were not the product of a well-developed theory of democracy. His understanding of democracy was more intuitive. He envisioned the democratic process as a largely self-regulating one. The analogy which immediately presents itself is to a free-market economy. In both cases, highly idealized visions are used to evaluate actual political and economic practices. Douglas's democratic intuition almost guarantees a wide disjunction between theory and practice. Real political actors will resist acting within a political process that provides them with no active role except to protect the integrity of the process, a role which appeared to gratify Douglas but few others. Still, his democratic intuition had broad appeal. It offered the comforting illusion of an impersonal, noncoercive democratic process. It also tapped a strong strain in American democratic thought which suspected statesmanship of being intrinsically undemocratic, even as it set aside the heretofore dominant strain in American democratic thought which considered statesmanship indispensable to well-run democracies.[28]

This vision of a largely self-regulating democratic process is pluralistic. In the previous section, we saw how Douglas's political thought was pluralistic in the sociological sense that he assumed neither the individual states nor the union were homogeneous republics. Now, we can see how his whole political thought pivots on the abandonment of the traditional republican assumption of social homogeneity. His passive view of statesmanship is a direct response to social diversity, to the belief that individual preferences are exceedingly diverse in modern democracies and that democratic statesmen should not act in such a way as to violate that diversity. The only remaining glue in Douglas's democratic society is the agreement of statesmen and citizens to its basic norms.

By the same token, Douglas's view of the relationship between democratic statesmen and their constituencies differs markedly from traditional republican views of the relationship between natural aristocrats and ordinary citizens. From his perspective, the traditional deference ordinary citizens show toward natural aristocrats is manifestly undemocratic. If anything, democratic statesmen should defer to their constituents. But, mostly, the entire relationship becomes attenuated. Douglas's democratic intuition evidences a long evolution from the founders' vision of a popular government of mixed aristocratic and democratic elements. The public good, at least as traditionally conceived, is another victim of his pluralistic understanding of democracy. It can only possess a negative meaning, as a social and political fabric which does not violate the autonomy of individual preferences. Civic virtue also takes on a largely negative meaning. Interpersonal comity refers to a disposition not to act in an obtrusive manner rather than to the participatory mode of citizenship favored by eighteenth-century republicans. The mutual indifference of citizens toward each others' preferences is a pluralistic substitute for disinterestedness, but the two qualities are hardly equivalent.

What remains of republicanism in Douglas's political thought is an important sociological argument against political consolidation and a certain ambivalence toward social diversity. He is sure that diversity is endemic to modern democracies, but he is not quite sure how a robust, democratic politics is possible in the face of it. The twentieth-century pluralists differ from Douglas in nothing so much as in believing that social diversity positively contributes to a robust, democratic politics.[29]

Racial Exclusions

Perhaps, however, I have unduly discounted the republican character of Douglas's political thought by not explicitly considering his public statements on slavery. Those statements seem very republican in two respects. First, they give priority to public over private liberty. Second, they justify racial slavery, or at least racial exclusions from the citizen body, on the basis of the alleged incapacity of blacks for self-government. Both arguments are, *prima facie,* republican arguments.

In fact, Douglas appears to place public liberty so far above private liberty that he could not even be a liberal democrat. He must, instead, be a majoritarian democrat who believes that the collective right of self-government is not limited by any private rights or other moral constraints. The slavery issue does seem to directly present the question of the limits on majority rule and Douglas does seem to answer that question in the negative.

> The great principle is the right of every community to judge and decide for itself, whether a thing is right or wrong, whether it would be good or evil for them to adopt it; and the right of free action, the right of free thought, the right of free judgment upon the question is dearer to every true American than any other under a free government . . . It is no answer to this argument to say that slavery is an evil, and hence should not be tolerated. You must allow the people to decide for themselves whether it is a good or an evil. (Johannsen, pp. 27–28 [Douglas])[30]

The reason we cannot take such statements at face value is that they were spoken in the context of discussing the question of the limits on the collective right of citizens to define the private rights of noncitizens; specifically, the collective right of white citizens to define the private rights of black noncitizens. Douglas's answer to that question need not prejudice his answer to the question of the limits on the collective right of white citizens to define the private rights of white citizens (or white noncitizens). His answer, therefore, may not be evidence of illiberalism or, more weakly construed, of liberal republicanism (as opposed to pluralism) so much as it is evidence of racism. According to Douglas, the American political tradition supports the unfettered liberty of white citizens to decide upon the putative rights of black noncitizens.[31]

The Constitution is not a slave document for Douglas, as it was for many of the abolitionists and proslavery apologists. It does not preclude black citizenship. However, it also does not prevent white majorities from treating black Americans very differently from white Americans. It leaves those decisions up to state majorities. They can grant blacks full citizenship rights. Or, they can keep blacks in servitude. Or, finally, they can emancipate blacks and grant them some private rights without making them full citizens—the policy actually favored by Douglas and Lincoln and common to most of the Northern states (see Johannsen, pp. 33, 46–47, 129, 266, 299 [Douglas]; pp. 52–53, 162, 197 [Lincoln]).[32]

Equally important, Douglas must show that the Declaration of Independence does not debar white majorities from freely determining what rights, if any, blacks possess. He, thus, claims that the signers of the Declaration did not intend the phrase, "all men are created equal," to refer to blacks.

> Now, I say to you, my fellow-citizens, that in my opinion, the signers of the Declaration had no reference to the negro whatever, when they declared all men to be created equal. They desired to express by that phrase white men, men of European birth, and European descent, and had no reference either to the negro, the savage Indians, the Fejee [*sic*], the Malay, or any other inferior and degraded race, when they spoke of the equality of men. (Johannsen, p. 128 [Douglas])[33]

Consequently, the rights of black Americans to "life, liberty, and the pursuit of happiness" are not "unalienable"; they, on the contrary, are completely subject to majority rule. But it is important to emphasize the converse. Douglas never denies the "truths" of the Declaration of Independence as they apply to white Americans. Whites are endowed with certain unalienable rights which are not completely subject to majority rule. For the most part, Douglas places the public liberty of white citizens far above the private liberty of black noncitizens. It is impossible to imagine him substituting "whites" for "negroes" in the statement that "I care more for the great principle of self-government . . . than I do for all the negroes in Christendom" (Johannsen, p. 326 [Douglas]). Such statements suggest that what was really central to Douglas's thinking about slavery was racism, not republicanism or illiberalism (except insofar as racism is itself illiberal).

Douglas's rationale for his disparate treatment of the rights of white and black Americans is racial inferiority. He claims that blacks constitute an inferior race incapable of self-government and that they, therefore, should never become United States citizens (see Johannsen, pp. 45, 127, 196, 216 [Douglas]). Again, this argument is, at first sight, evidence of republicanism. Blacks allegedly do not meet the rigorous standards of republican citizenship. However, the racial criterion of who can and who cannot meet those standards is immediately suspect, especially given that Douglas appears to have substantially lowered the standards for white Americans from civic virtue to mutual indifference. It is difficult to resist the conclusion that Douglas's racist attitudes prevented his discussion of the relation between majority rule and slavery from being an accurate representation of his views on the (abstract) relation between majority rule, private rights, and citizenship.

The ultimate irony of Douglas's discussion of majority rule and slavery emerges through his public interpretation of his recent opposition to the Lecompton constitution. During the last debate at Alton, Douglas lambasts his Democratic colleagues for sacrificing the white settlers' right of self-government to party expediency.

> Most of the men who denounced my course on the Lecompton question, objected to it not because I was not right, but because they thought it expedient at that time, for the sake of keeping the party together, to do wrong. I never knew the Democratic party to violate any one of its principles out of policy or expediency, that it did not pay the debt with sorrow. There is no safety or success for our party unless we always do right, and trust the consequences to God and the people. I chose not to depart from principle for the sake of expediency in the Lecompton question, and I never intend to do it on that or any other question. (Johannsen, p. 293 [Douglas])

This passage sounds like a ringing affirmation of democratic statesmanship against mere party expediency. Yet, it implies that the greatest acts of statesmanship are those which defend the rights of other citizens to decide for themselves whether members of "inferior" races shall enjoy any rights and which ones. Would not the greater acts of statesmanship be those which assert the unalienable rights of the members of all races and urge other citizens to fully recognize that fundamental equality?

Lincoln, of course, thought so. He did not believe popular sovereignty monopolized the question of right involved in the slavery issue. Accusing Douglas of a tragic lack of statesmanship, Lincoln both invited him to join the Republican Party in publicly denouncing slavery as an evil institution and attacked his failure to do so. Lincoln well knew that what principally separated them was not their personal preferences on the slavery issue but their willingness to act on those preferences in the face of strongly opposing ones. He also suggested that different understandings of democracy, alternately, underwrote or undermined that willingness.[34] Douglas could not help feeling that the rhetoric and policies of the Republican Party were prejudicial to the democratic privileges of their fellow citizens. For his part, Lincoln could not help feeling that a rudderless democracy would soon run aground.

8

Abraham Lincoln and the House United

Throughout his political career, first as a local Whig politician and then, reborn, as a Republican party leader, Abraham Lincoln never deviated from Webster's precept that liberty and union are one and inseparable. Indeed, his Solomonlike efforts to follow that precept when its two components seemed to be pulling so many other Americans apart became definitive of his political thought and action. Yet, his understanding of liberty and union differed from Webster's. They were not really following the same principle of action.[1]

Lincoln's understanding of liberty and union departed from Webster's in two important respects. First, he substituted a fairly loose, pluralistic definition of the union for Webster's tighter, more republican definition. Realistically, he could only emphasize the nationalist argument that diversity is a bond of union while deemphasizing the national-republican argument that diversity is superficial to deeper unities. Second, he substituted a more universalistic definition of liberty and union for Webster's essentially particularistic definition. For Lincoln, unlike Webster, the presence of slavery within the union seriously tarnished its world-historic mission for liberty. In his famous 1854 Peoria speech, which he quoted extensively during his debates with Douglas, Lincoln professed to "hate" the Democrats' declared indifference to the spread of slavery "because it deprives our republican example of its just influence in the world—enables the enemies of free institutions, with plausibility, to taunt us as hypocrites—causes the real friends of freedom to doubt our sincerity" (Johannsen, p. 50 [Lincoln]).[2]

Later in the Peoria speech, Lincoln claimed that, if alive, Webster (as well as Clay) would have joined him in opposition to Douglas's Kansas-Nebraska Act.[3] This claim is probably true. Webster, like Lincoln and unlike Douglas, would not have seen the Compromise of 1850, which all three men supported, as a precedent for the Kansas-Nebraska Act.[4] The conundrum here is whether Webster would have evolved in the same antislavery direction as Lincoln did and further apart from Douglas. Again, he probably would have. However, I doubt that he would have evolved as far as Lincoln did.[5] His principle of action would have remained "preserve the union while condemning slavery as much as possible," not Lincoln's "condemn slavery as much as possible while preserving the union." The difference in emphasis may seem subtle, but its historical significance was momentous. Ideologically, it inextricably linked the fulfillment of the American mission to the ultimate abolition of slavery. Politically, it favorably positioned the Republicans between the abolitionists and the "old-line" Whigs and Democrats in the North. In 1858, Lincoln elaborated the principle of action he and his party would continue to follow for the next five years, until the exigencies of civil war forced a change in policy to immediate emancipation.

> The Republican party think it [slavery] wrong—we think it is a moral, a social and a political wrong. We think it is a wrong not confining itself merely to the persons or the States where it exists, but that it is a wrong in its tendency, to say the least, that extends itself to the existence of the whole nation. Because we think it wrong, we propose a course of policy that shall deal with it as a wrong. We deal with it as with any other wrong, in so far as we can prevent its growing any larger, and so deal with it that in the run of time there may be some promise of an end to it. We have a due regard to the actual presence of it amongst us and the difficulties of getting rid of it in any satisfactory way, and all the Constitutional obligations thrown about it. I suppose that in reference both to its actual existence in the nation, and to our Constitutional obligations, we have no right at all to disturb it in the States where it exists, and we profess that we have no more inclination to disturb it than we have the right to do it . . . [But w]e also oppose it as an evil so far as it seeks to spread itself. We insist on the policy that shall restrict it to its present limits. (Johannsen, pp. 254–55 [Lincoln])[6]

Lincoln's *sine qua non* was a free-soil policy that would legally exclude the institution of slavery from all the new territories. "Free soil" was the one federal policy he considered constitutional and yet dealt with slavery as "a moral, a social and a political wrong."[7] This policy distinguished him from Webster but even more from Douglas. Although neither Webster nor Douglas could claim to have publicly supported free-soil policies (despite their private antislavery preferences), only Douglas sought to rally his party around a "declared indifference to the spread of slavery."[8] It, moreover, was Douglas's position to which Lincoln counterposed his party's free-soil policy in order to reassert the nation's commitment to equal liberty. Without that reassertion, he thought that the liberty of all Americans, black and white, was insecure, *and so was their union.*

The differences between Lincoln and Douglas on liberty and union were not the differences between Lincoln and Webster. The Lincoln-Douglas debates largely transcended the first- and second-generation debates between federal and national republicans. In some respects, Lincoln did fit into a common tradition with Webster and Publius, just as Douglas did with Calhoun and the Federal Farmer. Those two intellectual (sub)traditions, however, had undergone such extensive transformations over time that they no longer seemed to be the same traditions or that distinct from each other. For the most part, the Lincoln-Douglas debates involved a quarrel between two pluralists. This quarrel had three major aspects: (1) their different definitions of union; (2) their different analyses of American society; and (3) their different understandings of democracy.

First, Lincoln and Douglas disagreed over how much coherence is necessary in any modern pluralist society. In this respect, Lincoln's "House Divided" doctrine presents itself as a caution. It warns that even a pluralist society must be united on something more than on certain procedural principles. Such a society must also be united on certain substantive principles. Otherwise, Lincoln yielded to Douglas's pluralistic definition of the union.[9]

Second, Lincoln and Douglas differed in their empirical assessments of how much coherence actually existed in their own pluralistic society. Douglas argued that the union was not united, and could not in the foreseeable future be united, on the substantive principle Lincoln had most in mind—the principle that slavery is wrong. Because, according to Douglas, Americans fundamentally disagreed over that principle, they must be allowed to so disagree for the sake of their continued union. Lincoln adopted a quite different view of the matter. He argued not only that the union could be united on an antislavery principle but that it currently was so united. He assumed the existence of an underlying national consensus on the principle that slavery is a necessary evil. He also assumed that a union which was as divided as Douglas contended the American union was would soon fall apart of its own accord anyway.[10]

Given their different beliefs about the nature of pluralist societies and about the nature of their own pluralistic society, it is hardly surprising that Lincoln argued for a national decision on the slavery issue and Douglas for a series of local decisions. Their debates, though, were not primarily about federalism, certainly not primarily about federal republicanism. Both men were strong nationalists and neither man contested the constitutional right of states, or prospective states, to decide the slavery issue for themselves. Only with respect to the territories and only on the level of public sentiment did Lincoln insist on a national decision. This insistence pointed less to a different understanding of states' rights than to a different understanding of how pluralist societies best function and, more specifically, of how pluralist democracies best function.[11]

Third, then, Lincoln and Douglas held different ideas of democracy and democratic statesmanship. For Lincoln, the essence of democracy was not the mutual respect and self-restraint of citizens and statesmen toward each others' preferences on controversial political issues but rather their efforts to help each other make the right decisions on those issues. Lincoln rejected Douglas's democratic intuition for a more dynamic and teleological view of democracy. This difference was fundamental, as can be demonstrated by continuing to work through Lincoln's "House Divided" doctrine.

According to that doctrine, the union must, above all else, be united on the slavery issue because it is the only issue that threatens to tear the union apart. The doctrine, however, offers a choice between two possibilities: either "freedom national" or "slavery national." Lincoln, of course, strongly favored the first possibility.[12] But why? How did he propose to act (or not act) on that preference? And did he really think both possibilities were equally open to the future?

Lincoln's beliefs about the nature of pluralist societies seem to leave open both possibilities. The union must be united on some substantive principle(s), but presumably the principle that slavery is right (or not wrong) qualifies equally with the principle that it is wrong. What those beliefs explicitly exclude is a third possibility: Douglas's possibility of an indefinitely divided house based on a mutual indifference to the rightness or wrongness of slavery. Herein lay Lincoln's genius in framing the issue as he did. He realized that if he could eliminate Douglas's middle position, Douglas himself, as well as the bulk of the northern Democracy, would have no choice but "freedom national."[13] Lincoln even thought that once the issue was clearly drawn between "freedom national" and "slavery national," the nation, as a whole, could not fail to choose the first possibility. He felt "freedom national" was the only possibility which was really open to the future. In retrospect, he obviously underestimated the strength of those groups in the South who would choose yet a fourth possibility, a possibility which the "House Divided" doctrine discounted as the default position no one would ever willingly choose. This was the possibility of disunion.[14]

Nonetheless, Lincoln did not favor "freedom national" just because he thought it was the only realistic possibility for the future. His "House Divided" doctrine expressed a genuine concern for a possible erosion of public antislavery sentiments and his advocacy of the opposite scenario clearly was influenced by his own antislavery sentiments. Douglas, however, shared similar sentiments and he did not advocate the same scenario. This disjunction between their private preferences and public attitudes must be explained.

One line of investigation would examine how Lincoln's and Douglas's different understandings of pluralism, nationalism, and social realism required them to act on the slavery issue. The explanation here would be that, notwithstanding their similar antislavery preferences, those three factors impelled Douglas to take a neutral position on the issue, while impelling Lincoln to take an explicit antislavery position. As I have already suggested, this explanation is important but ultimately unsatisfying. A second line of investigation would pursue how Lincoln's and Douglas's antislavery preferences did differ. The explanation would now focus on how, and in what ways, Lincoln's antislavery preferences were stronger than Douglas's as the principal reason he was more likely to act on those preferences. This explanation is also important, especially to the extent that Lincoln, unlike Douglas, saw slavery as destructive of liberal society. Yet, it still seems to miss the crux of their differences. A third line of investigation is, therefore, necessary. It would assume that Lincoln and Douglas did not differ so much in their antislavery preferences as in how they proposed to act on those preferences and that their different understandings of democracy, more than of pluralism, nationalism, and social realism, determined their different principles of action on the slavery issue. In Douglas's case, his understanding of democracy consigned him to virtual silence about his own antislavery preferences. Conversely, Lincoln's understanding of democracy enjoined him to bear witness to his antislavery preferences in an attempt to prejudice the future toward "freedom national." The three sections of this chapter will be organized around these three lines of investigation.

Before turning to the body of the chapter, I, however, would like to pause to explicitly consider the republican-revisionist interpretation of Lincoln's political thought.[15] Admittedly, his political thought does reflect many traditional republican ideas. This republican strand is most evident in his understanding of democracy, much more so than in the case of Douglas. His ideas on democratic statesmanship harken back to the eighteenth-century vision of a disinterested natural aristocracy. Lincoln, nevertheless, represented a new type of democratic statesman, one who was "a man of the people" and one who emphasized the linkages between his own political rhetoric and public opinion. He, like Douglas, better fits into a pluralist than a republican paradigm of politics. The efforts of the republican revisionists to fit Lincoln and his party into the latter paradigm are particularly unconvincing, though, in their errors, not unilluminating.

The revisionists fasten on two features of Lincoln's political rhetoric. First, they stress his conspiracy theory. Most of the "House Divided" speech is devoted to exposing a conspiracy between "Stephen

[Douglas], Franklin [Pierce], Roger [Taney] and James [Buchanan]" to impose the institution of slavery on the whole nation through a second "Dred Scott" decision which would forbid states, as well as territories, from outlawing slavery within their boundaries (see Johannsen, pp. 14–20 [Lincoln]). This charge mirrors Republican attacks on a Southern "slavocracy" for capturing the federal government with the ultimate aim of depriving other whites of their basic liberties. It also echoes earlier American statesmen in their nagging fears for the future of the republic, in their persistent suspicions that the republic is being subverted by conspiratorial "court" parties, and in their periodic attempts to organize "country" parties to combat these plots against the republic. Second, the revisionists stress Republican defenses of the party's "free-soil" policy on grounds that it is necessary to protect the vitality of a "free-labor" system. Lincoln offers such a defense during the Alton debate.

> Now irrespective of the moral aspect of this question as to whether there is a right or wrong in enslaving a negro, I am still in favor of our new Territories being in such a condition that white men may find a home . . . where they can settle upon new soil and better their condition in life. (Johannsen, p. 316 [Lincoln])

This passage suggests that Lincoln accepted the following account: Because free and slave labor do not mix, and because the older parts of the union are already "overcivilized," the new territories must be kept free of slaves in order to preserve a republic which is predominantly peopled by independent artisans and yeoman farmers.

The problem with putting so much weight on Lincoln's conspiracy theory is that he did not believe the actions of the Democratic leadership of "Stephen, Franklin, Roger, and James" constituted a conspiracy in the usual sense of the term.[16] His subsequent exchanges with "Stephen" during the course of the campaign reveal he did not believe that their actions were the result of a secret pact between them to nationalize the institution of slavery, or even that it was the intention of any one of them to do so. He, on the contrary, only viewed their actions as tending in that direction. In fact, his conspiracy theory provides stronger evidence of an implicit critique of Douglas's democratic intuition for discouraging serious reflection about the tendencies of political actions and events than it does of any traditional republican fears of conspiratorial parties. Lincoln still could have thought Douglas was an unwitting agent of a Southern slavocracy that did intend to nationalize slavery. His campaign speeches, though, are devoid of any such references to a slavocracy and are remarkably charitable to the South. He clearly thought most Southerners, just like most Northerners, were not proslavery so much as they were constrained by a nexus of interests which made abolishing the institution seem impossible to them.[17] Nor did Lincoln deny the legitimacy of the Democratic Party, however much he might have considered their policies misguided. Furthermore, he, like Douglas, was a loyal member of his own party—which is counterevidence to a republican interpretation.[18]

The revisionists' reliance on Lincoln's "free-soil/free-labor" argument is equally misplaced. They misconstrue the primary thrust of the free-soil policy, which was symbolic, not causal. Again and again, Lincoln returns to the argument that the free-soil policy will put slavery "in the course of ultimate extinction" (see Johannsen, pp. 132, 155, 199–200, 277–78, 310 [Lincoln]). The argument, however, is not a causal one. It is not that the free-soil policy will directly (even if gradually) eradicate slavery in America, for example, by isolating and, thus, economically strangling the institution in the Southern states.[19] No, the argument is that the free-soil policy will indirectly (and still very gradually) lead to the end of slavery by reasserting the nation's commitment to equal liberty. It is doubtful if Lincoln even felt that the free-soil policy was necessary to keeping slavery out of the new territories. Douglas argued that it was not; that nature, or, more precisely, nature working through human self-interest, had reserved the territories for freedom. While Lincoln rejected that argument, he also believed that the moral, rather than the legal, effects of the free-soil policy would be decisive in excluding slavery from the territories (cf. Johannsen, pp. 146–47 [Lincoln]; pp. 160–61, 269–70, 328–29 [Douglas]).[20] Similarly, the threat of a second "Dred Scott" decision did not trouble him so much as the threat of a tragic slippage in public antislavery sentiments, a slippage which Douglas and his "co-conspirators" were fostering and which would make such a court decision possible. In making these points, I do not mean to deny that Lincoln thought the free-soil policy would have some concrete, causal consequences. But I do mean to claim that he considered the effects it would have on public sentiments the more important ones and that the revisionists mistakenly reverse that emphasis.

Douglas preceded the revisionists in focusing on the causal dimensions of the free-soil policy. As we have seen, he accused Lincoln of literally trying to starve slavery out of the South. If the free-soil policy had a more symbolic significance to Lincoln, then he would, of course, have been less open to this criticism. He, however, would have been more open to the criticism that he had no blueprint for ending slavery in America. Lincoln himself admitted that he had no blueprint. He did suggest gradual, compensated emancipation and colonization as partial solutions. Yet, he did not appear to have much confidence in such schemes, which was probably why he stretched the time frame for eradicating slavery to a hundred years (see Johannsen, pp. 55, 200 [Lincoln]).[21] The abolitionists had a blueprint, but it was one Lincoln and most other Americans, North and South, vehemently opposed. The abolitionists, moreover, had no blueprint for a post-emancipation society.[22] For Lincoln, the free-soil policy was a matter of putting first things first; first, the nation's commitment to equal liberty had to be reasserted and then the question of how and when slavery would be abolished would inevitably move to the forefront of political debate.

The broader conclusion the revisionists draw from Lincoln's "free-soil/free-labor" argument is also problematic. He clearly was not committed to a traditional republican economy of independent artisans and yeomen farmers. To the contrary, he was committed to a free-labor system that was becoming increasingly commercial and industrial, and he did not betray any nostalgia for the simpler economic order of the early republic. Lincoln's economic outlook, like Douglas's, was thoroughly progressive—again, counterevidence to a republican interpretation.[23]

Above everything else, the republican revisionists unfairly denigrate the moral side of Lincoln's politics by contending that the free-soil policy was primarily intended to benefit Northern whites and only incidentally black slaves. They aim to debunk and their interpretation dovetails with "ulterior-motive" accounts which stress Lincoln's consuming political ambition.[24] To be politically successful, he, after all, had to appeal to racist white voters in Illinois and other Northern states. Lincoln certainly was an ambitious man. This view, though, fails to distinguish him from Douglas or any other ambitious Northern politician at the time.[25] It does not explain why Lincoln's and Douglas's ambitions took such different political paths. To this task, I now turn.

"An Apple of Discord"

In this section of the chapter, I will try to explain the different positions Lincoln and Douglas took on the slavery issue in terms of their different readings of the current exigencies of the union. These different readings provoked a number of empirical (and quasi-empirical) disputes which, while important, fail to adequately account for the directionality of Lincoln's "House Divided" doctrine as compared to the neutrality of Douglas's popular-sovereignty doctrine. Such factors as nationalism, pluralism, consensus, social realism, filiopiety, and history explain why Lincoln insisted on a united house instead of a divided house but not why he insisted on "freedom national" instead of "slavery national."

Nationalism and Pluralism

As we saw in chapter 7, Douglas claims that Lincoln's "House Divided" doctrine is simply not true. In order to preserve the union, national uniformity is no more necessary in the laws regulating race relations than it is in the laws regulating cranberry orchards. Besides, diversity has important economic (and noneconomic) benefits. During their first debate at Ottawa, Lincoln responds by arguing that Douglas has overlooked the fundamental difference between the two cases.

> When he undertakes to say that because I think this nation, so far as the question of slavery is concerned, will all become one thing or all the other, I am in favor of bringing about a dead uniformity in the various States, in all their institutions, he argues erroneously. The great variety of the local institutions in the States, springing from differences in the soil, differences in the face of the country, and in the climate, are bonds of Union. They do not make "a house divided against itself," but they make a house united. If they produce in one section of the country what is called for by the wants of another section, and this other section can supply the wants of the first, they are not matters of discord but bonds of union, true bonds of union. But can this question of slavery be considered as among these varieties in the institutions of the country? I leave it to you to say whether, in the history of our Government, this institution of slavery has not always failed to be a bond of union, and, on the contrary, been an apple of discord, and an element of division in the house. (Johannsen, p. 54 [Lincoln])

In subsequent debates, Lincoln details how the slavery issue, unlike other "merely" economic issues, has been "an apple of discord." Douglas contests the evidence (cf. Johannsen, pp. 136, 198–200, 236, 313–14 [Lincoln]; pp. 129–31, 288 [Douglas]). This dispute seems to be simply a factual dispute over whether the union can endure its divisions on the slavery issue.

But, of course, factual disputes are never simply factual disputes. First, there is the matter of the disputants' common nationalism. Lincoln and Douglas implicitly agree that if the union cannot endure its divisions on the slavery issue then those divisions must somehow be overcome.[26] Lincoln, however, would balk at agreeing to the converse, that if the union can endure such divisions then the attempt should *not* be made to overcome them. That proposition is one side of Douglas's nationalist ethic; the other side is the counterfactual that the union cannot survive the attempt to overcome its divisions on the slavery issue. Still, the heart of the dispute is the exigencies of the union. Douglas's popular-sovereignty doctrine, equally to Lincoln's "House Divided" doctrine, is defended as the best means to preserving the union. Federalism just does not seem that important to either of them except insofar as it affects the viability of the union. Both Douglas's more positive commitment to states' rights and Lincoln's more negative commitment seem directed to that end. In Lincoln's case, he casts his solicitude toward states' rights as one of the practical necessities he must accept on the slavery issue for the sake of the union (see Johannsen, pp. 52, 131–32, 221, 278, 315 [Lincoln]).

Second, there is the disputants' common pluralism. Lincoln and Douglas agree that modern societies can endure tremendous diversity and that, in general, diversity is socially beneficial. Lincoln establishes this area of agreement in his response to Douglas's equation of slave laws to cranberry laws. The definition of the union underlying that response is highly pluralistic. Lincoln's distinction between the two cases, thus, constitutes a qualification. Even a highly pluralistic society cannot endure divisions on such a fundamental issue as the rightness or wrongness of human bondage. The society, as a whole, must make a decision on the merits of the slavery issue. Douglas disagrees. A pluralistic society can equally endure divisions on slave laws and cranberry laws. It only requires a consensus on certain procedural principles which permit local communities to decide those issues for themselves free from outside interference. Yet, the dispute is not primarily a dispute between a national and a federal republican over the size and coherence of the decision-units. It is primarily a dispute between two pluralists over the exigencies of a particular pluralistic society.

Nationalism and pluralism, though, only explain why Lincoln rejected Douglas's indefinitely divided house. Those two factors alone do not explain why he chose a house united on the wrongness of slavery instead of one united on the rightness of slavery. We must continue to unpack his "House Divided" doctrine to determine if it offers us any guidance as to why he chose the first possibility.

Consensus and Social Realism

The initial answer has to be that Lincoln believes the nation has already made a choice in favor of the principle that slavery is wrong. He is confident that whenever the nation is clearly presented with a choice between that principle and the opposing principle that slavery is right, it will unhesitatingly choose the former. Accordingly, he claims that "whenever we can get the question distinctly stated—can get all these men who believe that slavery is in some of these respects wrong, to stand and act with us in treating it as a wrong—then, and not till then, I think we will in some way come to an end of this slavery agitation" (Johannsen, p. 257 [Lincoln]).[27]

By all appearances, Lincoln intends to include the South in this national antislavery consensus (although it is also a consensus which believes slavery is presently a necessary evil). His view of the Southern people is that "[i]f slavery did not now exist among them, they would not introduce it" (Johannsen, p. 51 [Lincoln]).[28]

Lincoln invokes a certain sense of social realism in choosing "freedom national" over "slavery national," just as he did in rejecting a divided house. Given the national distribution of preferences on the slavery issue, "freedom national" is the most realistic choice. Again, a disagreement of fact emerges. Douglas does not see the same distribution of preferences that Lincoln does. He does not think a latent antislavery consensus exists in America which Lincoln and other statesmen only need to cultivate in order to solve the current sectional crisis. He, rather, sees pockets of widely divergent preferences—antislavery, proslavery, and many shades in between—which make a divided house the most realistic choice (see Johannsen, pp. 33, 47, 129, 216, 299–300 [Douglas]).

Nonetheless, Lincoln's choice could not have been solely, or even primarily, based on grounds of social realism, for that assumes he did not take the other possibility seriously. He, however, did take the possibility of "slavery national" seriously. Indeed, he thought Douglas, among others, was preparing the way for "slavery national" by obscuring the issue between it and "freedom national" (see Johannsen, pp. 257, 320 [Lincoln]). Lincoln, conversely, would meet the crisis by clearly presenting the issue between those alternative futures through his "House Divided" doctrine. He would not have acted in such a manner if he had been indifferent between those two futures, nor if he had not wished to prejudice the future toward "freedom national" and if he had not felt that future was somehow at risk.

Filiopiety

Lincoln's belief that the founders originally created a house united on antislavery principles is perhaps a more significant factor in giving directionality to his "House Divided" doctrine. Douglas, of course, believes that the founders created a divided house. The two men, nevertheless, agree that the principles and policies of the founding fathers carry great weight. They "merely" disagree about the nature of those principles and policies. This disagreement defines another set of factual disputes (cf.

Johannsen, pp. 56, 132–33, 278, 312, 320 [Lincoln]; pp. 32, 44, 125, 197, 218, 288, 326–27 [Douglas]).[29]

In rebutting Douglas's claim that the founders created a divided house, Lincoln notes that even if the house had then been divided over the slavery issue, the founders had not created that situation; they had only accepted it as something they could not change immediately (see Johannsen, pp. 277, 303–4, 311–12 [Lincoln]).[30] Lincoln, however, also denies that the house was then divided, at least not in the relevant sense of the term. What makes for a divided house is not the presence of the institution of slavery in some states and its absence in others but the existence of deep differences of opinion between the states over the rightness or wrongness of the institution. Lincoln contends that in the latter sense the union was not divided over the slavery issue during the Revolutionary era, any more than it is in the late 1850s.[31] He, thus, answers Douglas's query about how the union could have survived so long divided on the slavery issue, if, as he insists, it must be united on the issue, by denying the premise.

> . . . and when the Judge reminds me that . . . the institution of slavery has existed for eighty years in some States, and yet it does not exist in some others, I agree to the fact, and I account for it by looking at the position in which our fathers originally placed it—by restricting it from the new Territories where it had not gone, and legislating to cut off its source by the abrogation of the slave trade thus putting the seal of legislation against its spread. The public mind did rest in the belief that it was in the course of ultimate extinction. (Johannsen, pp. 54–55 [Lincoln])[32]

The question of the founders' intentions toward slavery goes beyond the question of whether they created a divided house. It also concerns their intentions toward the future of the institution. Lincoln argues that the founders looked forward to the eventual end of slavery in the United States and enacted a set of policies designed to promote that goal. Those policies included the constitutional clause authorizing Congress to prohibit the importation of slaves into the country after 1808 and the Northwest Ordinance of 1787, which outlawed slavery in the Ohio territories. In Lincoln's eyes, those two policies prove that the founders looked forward to the end of slavery.

> Let me ask why they made provision that the source of slavery—the African slave-trade—should be cut off at the end of twenty years? Why did they make provision that in all the new territory we owned at that time, slavery should be forever inhibited? Why stop its spread in one direction and cut off its source in another, if they did not look to its being placed in the course of ultimate extinction? (Johannsen, p. 310 [Lincoln])[33]

Lincoln, therefore, claims the founding fathers as the original free-soilers. Douglas does not refute Lincoln's evidence so much as refer it to a larger context, to the fact that the Constitution reserves the question of slavery to the states. Lincoln is, again, very (overly?) solicitous of that fact. He, however, interprets it quite differently. For him, it is not evidence that the founders were committed to Douglas's popular-sovereignty doctrine, nor that they were divided over the long-term fate of slavery. It, rather, confirms that they thought the abolition of slavery anytime in the near future was not feasible. And yet, according to Lincoln, they did what they could to place the institution "in the course of ultimate extinction"; they would have been very surprised to find it still an ongoing concern in 1858 (see Johannsen, pp. 132–33, 278, 320 [Lincoln]).[34] Douglas can hardly deny the latter point. He can only weakly reply that the founders would have been surprised by many features of mid-nineteenth-century American life, such as telegraphs and railroads (see Johannsen, p. 327 [Douglas]). Douglas would then place slavery on "the cotton-gin basis," as if the founders' expectations about the fate of the institution had simply constituted a scientific prediction which had gone awry because it failed to anticipate future inventions (see Johannsen, p. 320 [Lincoln]).

The Lincoln-Douglas dispute over the meaning of the Constitution, though, pales before their dispute over the meaning of the Declaration of Independence.[35] Lincoln views the Declaration as further evidence of the founders' antislavery intentions as well as of how those intentions were consistent with the continued existence of the institution of slavery.

As we have seen, Douglas repeatedly denies that the founders intended the central claim of the Declaration of Independence to refer to blacks. This denial truly outrages Lincoln. He insists that the founders intended to include blacks in the claim that "all men are created equal" and that until recently no one doubted that they did.

> I assert that Judge Douglas and all his friends may search the whole records of the country, and it will be a matter of great astonishment to me if they shall be able to find that one human being three years ago had ever uttered the astounding sentiment that the term "all men" in the Declaration did not include the negro . . . I believe the first man who ever said it was Chief Justice Taney in the Dred Scott case, and the next to him was our friend, Stephen A. Douglas. And now it has become the catch-word of the entire party. (Johannsen, pp. 304–5 [Lincoln])[36]

In support of his view, Douglas argues that if the founders had intended the claim that "all men are created equal" to refer to blacks, then they would have been hypocrites for not immediately freeing their slaves (see Johannsen, pp. 128, 215–16 [Douglas]). Lincoln suggests that Douglas, once again, has misinterpreted the evidence. The Declaration of Independence was not a public-policy statement but "a standard maxim for free society." It established the long-term goal of equal liberty for all Americans, but it also implicitly recognized that such a society is unattainable; not in 1776; not in 1858; not in 1993. Quoting from a speech he delivered the previous year in reaction to the *Dred Scott* decision, Lincoln contends that in the Declaration the founders

> meant to set up a standard maxim for free society which should be familiar to all: constantly looked to, constantly labored for, and even, though never perfectly attained, constantly approximated, and thereby constantly spreading and deepening its influence and augmenting the happiness and value of life to all people, of all colors, every where. (Johannsen, p. 304 [Lincoln])[37]

Lincoln's filiopiety goes quite deep. He not only sees himself as pursuing the same principles and policies as the founders, but he envisions himself in the same situation as the founders. The Declaration of Independence remains a standard maxim for his society. Immediate abolition still seems impossible, both because of the legal barriers, now including the founders' own Constitution, and because of the political dilemmas, such as how to preserve a union with former slaveholders and their freed slaves. And, all things considered, "free soil" continues to be the best policy.

Yet, Lincoln's filiopiety does not adequately account for his commitment to "freedom national." In the first place, he is well aware that his situation is not exactly the same as the founders' situation. He fears that recent events have disrupted the seemingly changeless context of the slavery issue in America, thus forcing him to respond with principles and policies which consciously go beyond the founders' legacy. As compared to them, he does advocate more explicit free-soil policies and he does more explicitly condemn the institution as a moral, social, and political wrong. In the second place, Douglas appeals to the same legacy to justify a very different set of principles and policies. We might think that Lincoln has the better of the argument on the nature of the founders' legacy, but even he must admit that, properly qualified, Douglas's popular-sovereignty doctrine forms *part* of that legacy (see Johannsen, pp. 131–32, 221, 315 [Lincoln]). The "filiopiety" explanation, finally, does not tell us why (or how) he and Douglas could derive such different principles and policies from their common patrimony.[38]

History

A somewhat broader explanation is that Lincoln viewed history as the arbiter of his "House Divided" doctrine. Here, the explanation is that he was not (merely) fixated on the teachings of the founding fathers but that he saw history, in a more general and abstract sense, as the principal source of human wisdom. This explanation obviously suffers from some of the same difficulties as the "filiopiety" explanation. However, it is undeniable that both Lincoln and Douglas looked to the past for guidance and found it very usable. Both discovered an apparently unbroken stream of precedents for their positions, even if they, again, disputed the nature of those precedents and even if Lincoln, at least, was disconcerted by the recent turn of events.

Paradoxically, Douglas argues that the slavery issue has not really been an apple of discord throughout American history, despite believing that the union has been divided on the issue, while Lincoln defends the converse position. They, then, disagree about the causes of the previous crises of the union. Lincoln claims that, at bottom, they have all been caused by the slavery issue; Douglas rejects that claim (cf. Johannsen, pp. 136–37, 199, 314 [Lincoln]; pp. 325–26 [Douglas]).[39] They also disagree about the policies their predecessors adopted to help solve those crises. Douglas insists that prior federal policies on slavery were *quid pro quos* that confirmed the divided nature of the house. Lincoln, in contrast, insists that those policies reflected a long-settled national commitment against the spread of slavery that has kept the institution "in the course of ultimate extinction" and maintained the united nature of the house (cf. Johannsen, pp. 54–55, 132, 277–78, 312 [Lincoln]; pp. 37–38, 187, 296, 325 [Douglas]).[40]

Not surprisingly, Lincoln and Douglas also offer different interpretations of the roots of the current crisis of the union. For Douglas, those roots lie in the attempt to unite a divided house; in policy terms, the Republican crusade to legally prohibit slavery in all the new territories. He contends that the current crisis, which, he admits, does centrally involve the slavery issue, has been trumped up by the Republican Party to serve its narrow sectional and partisan purposes (see Johannsen, pp. 118, 211, 300 [Douglas]). For Lincoln, the roots of the crisis lie in the attempt to divide a united house; in policy terms, the Democrats' Kansas-Nebraska Act and *Dred Scott* decision, which, by repealing the Missouri Compromise, threaten to reverse the free-soil thrust of past federal policies. He, thus, shifts the blame for the crisis, charging it to the Democratic effort to expand slave territory (see Johannsen, pp. 136, 236, 313–14 [Lincoln]).

To Lincoln's horror, this effort seems to be succeeding. He claims that Douglas "and those acting with him, have placed that institution on a new basis, which looks to the perpetuity and nationalization of slavery" (Johannsen, p. 55 [Lincoln]). Underlying the Democrats' success, Lincoln detects a deterioration in public opinion in a more proslavery direction.[41] This deterioration requires Republicans, above all else, to shore up the antislavery consensus, once more going beyond their predecessors in explicitly condemning slavery and in enacting explicit free-soil policies. These principles and policies are necessary to restore the public mind to its belief that slavery is in the course of ultimate extinction.

Yet, that same shift in public opinion also suggests that history cannot be the consummate guide for Lincoln; that he must possess standards outside history without which he could not judge reversals as reversals. He and Douglas, moreover, interpret history in considerably different ways. This explanation, again, does not tell us how and why. Although history was undoubtedly important to each man, neither appears to have been especially bound by it.

In sum, each of the four preceding explanations of Lincoln's "House Divided" doctrine adds something to our understanding of that doctrine without, either singly or collectively, adequately explaining the directionality of, in truth, the moral force behind it. I will next consider four other explanations that, at least ostensibly, have a more normative or ethical focus. They, therefore, seem more commensurate to explaining Lincoln's antislavery politics and how it differed from Douglas's "I don't care" politics.

Equal Liberty

In the previous section of this chapter, I assumed that Lincoln and Douglas shared antislavery preferences which were similar in strength and character (or that it did not matter how they might have differed in those respects). I then tried to explain their different positions on the issue using factors exogenous to their preferences on the slavery issue. In this section, I will open up the assumption that they shared similar preferences on the issue and explore whether humanitarianism, Protestantism, republicanism, and liberalism identify any differences in the strength and character of their preferences. Each of these four factors defines the wrong of slavery in a different way. Our task, in part, will be to discover which factor was the most important one in defining the wrong of slavery for Lincoln. At the end of this discussion, however, we still will not have adequately explained Lincoln's position on the slavery issue and how it differed from Douglas's position. This is because the relevant difference was not their preferences on the issue (nor such exogenous factors as nationalism) but how they proposed to act on those preferences.

Humanitarianism

By humanitarianism, I mean the disposition of members of one race to think of members of other races as equal human beings. Lincoln and Douglas both thought of blacks as human beings but beyond that basic agreement they differed as to how equal they thought blacks were to whites. Looking backward from the perspective of our late twentieth-century racial attitudes, Lincoln was the more progressive of the two, yet he still was in certain respects a racist.[42]

We must be especially careful in examining racial attitudes to distinguish private belief from public expression. For both Lincoln and Douglas, the public expression was probably more racist than the private belief because of the prevalence of racist attitudes among their audiences. We might simply assume that insofar as they addressed the same audiences during their senatorial debates and insofar as Lincoln's statements on race were more humanitarian than Douglas's, he was the more humanitarian of the two. There, however, were a number of complicating factors. Their audiences were only physically the same; politically, they were divided in two. Douglas's Democratic "half" was undoubtedly the more racist half.[43] Lincoln and Douglas also made different assessments of the risks involved to their careers, parties, and nation in making strong humanitarian statements. In addition, their different understandings of democracy also counseled greater or lesser caution in making such statements. Nonetheless, the evidence does clearly suggest that Lincoln held more humanitarian attitudes.

Lincoln carefully distinguishes the senses in which he does and does not believe blacks are (or should be) equal to whites. He, once again, uses the Declaration of Independence as a fulcrum for making the desired distinctions, as if his distinctions were the founders' distinctions.

According to Lincoln, the authors of the Declaration of Independence intended to assert that "all men are created equal" only in certain fundamental respects.

> I think the authors of that notable instrument intended to include all men, but they did not mean to declare all men equal in all respects. They did not mean to say all men were equal in color, size, intellect, moral development or social capacity. They defined with tolerable distinctness in what they did consider all men created equal—equal in certain inalienable rights, among which are life, liberty, and the pursuit of happiness. (Johannsen, p. 304 [Lincoln])[44]

Earlier in the debates, Lincoln had stated this view in his own voice. He then deduced from it a demand against the continued existence of slavery and for a minimal type of social equality.

> I agree with Judge Douglas he [the negro] is not my equal in many respects—certainly not in color, perhaps not in moral or intellectual endowment. But in the right to eat the bread, without the leave of anybody else, which his own hand earns, he is my equal and the equal of Judge Douglas, and the equal of every living man. (Johannsen, p. 53 [Lincoln])

Lincoln, finally, specifies the types of social and political equality he does *not* favor.

> I will say then that I am not, nor ever have been, in favor of bringing about in any way the social and political equality of the white and black races—that I am not nor ever have been in favor of making voters or jurors of negroes, nor of qualifying them to hold office, nor to intermarry with white people; and I will say in addition to this that there is a physical difference between the white and black races which I believe will forever forbid the two races living together on terms of social and political equality. And inasmuch as they cannot so live, while they do remain together there must be the position of superior and inferior, and I as much as any other man am in favor of having the superior position assigned to the white race. (Johannsen, p. 162 [Lincoln])

This passage from the Jonesboro debate seems unequivocal in two respects. One, Lincoln does not believe in the possibility of a biracial society. Two, given that belief, he strongly prefers the continued social and political dominance of his own race. Lincoln is racist in those two respects.[45]

We, nevertheless, should also note how Lincoln hedges his statements on racial inequality.[46] He says that the black man is not his equal, "*certainly* not in color, *perhaps* not in moral or intellectual endowment." Skin color is, again, his referent when he says that "there is a physical difference between the white and black races" which precludes the possibility of a biracial society. Lincoln seems to be underscoring the general intransigence of racial prejudice rather than confessing to his own racial prejudices. He certainly does not appear to believe that blacks are naturally unequal to whites in any way which would brand them as the inferior race. Yet, the most telling passage in the debates regarding Lincoln's views on racial (in)equality is a long paragraph from the Peoria speech, which he quotes at Ottawa.

Lincoln opens this paragraph by expressing sympathy for the South's predicament, a predicament a wizened Thomas Jefferson likened to "holding a wolf by the ears" at the time of the Missouri controversy.[47]

How can the Southern slaveholders free their slaves if a biracial society is impossible? Lincoln "surely will not blame them for not doing what I should not know how to do myself." This predicament makes colonization schemes attractive, but Lincoln, a long-time supporter of such schemes, acknowledges that they are impractical on a large scale.[48] The next option is to free the slaves and keep them as an underclass. Lincoln thinks this option is preferable to keeping them as slaves but not by much.[49] He "would not hold one in slavery at any rate; yet the point is not clear enough to me to denounce people upon." The final option is to free the slaves and accord them equal rights to white citizens. Lincoln finds this option unattractive and, if not unattractive, impractical. His "own feelings will not admit of this; and if mine would, we well know that those of the great mass of white people will not." In the end, though, Lincoln does know how he would act and, despite his characteristically charitable language, he does implicitly blame the Southern states for not acting in that manner. He closes the paragraph by suggesting that "systems of gradual emancipation might be adopted; but for their tardiness in this, I will not undertake to judge our brethren of the South" (see Johannsen, pp. 51–52 [Lincoln]). For Lincoln, the South's predicament is not exactly Jefferson's "wolf by the ears." It is more like making sure that everyone is following the same roadmap on a tortuous, cross-country journey. Lincoln insists on sustaining progress toward a society without racial slavery even though no consensus exists on what form such a society will take.[50]

Lincoln's expressed purpose for quoting from his Peoria speech at Ottawa is to rebut a charge which is emerging as the main theme of Douglas's campaign. This charge is that Lincoln is a "black" (radical) Republican, a fellow traveler of the despised abolitionists (see Johannsen, pp. 39–43 [Douglas]). According to Douglas, Lincoln favors full social and political equality for blacks and, therefore, is out of step with the average "white" (racist) voter of Illinois. Ironically, no part of the debates lends more support to the charge. Lincoln clearly was not an abolitionist. He did not advocate immediate, universal emancipation, nor did he propose full equality for the freed slaves.[51] He, however, resembled the abolitionists in insisting that universal emancipation remained his nation's long-term goal, which it must strive to attain "as fast as circumstances should permit" (Johannsen, p. 304 [Lincoln]). He also resembled the abolitionists in being a strong humanitarian. Like them, he stressed how slavery denied the humanity of the slaves by treating them merely as property. Not surprisingly, Douglas kept harping on this "guilt by association" charge in spite of Lincoln's numerous disclaimers (see Johannsen, pp. 97–104, 118–24, 189–96 [Douglas]).[52]

The logic of Lincoln's response is that Douglas, unwittingly or not, fails to see the middle position he has carefully staked out on racial equality. His favorite retort is that "I do not understand that because I do not want a negro women for a slave I must necessarily want her for a wife" (Johannsen, pp. 162–63 [Lincoln]).[53] He also claims that the abolitionists have no place in the Republican Party, though neither do those who think slavery is not wrong (see Johannsen, pp. 255, 317 [Lincoln]). He sometimes puts Douglas and the whole Democratic Party in the latter category to the extent that they deny the humanity of the slaves. He explicitly accuses Douglas of reducing "the whole question of slavery to one of a mere right of property" (Johannsen, p. 20 [Lincoln]).[54] Obviously, he believes that the average Illinois voter is not as "white" as Douglas does and that it is he, not Douglas, who is in step with the electorate.

Lincoln, in effect, accuses Douglas of being too racist for the voters of Illinois. Douglas was more of a racist than Lincoln was. Douglas, not Lincoln, claims that blacks constitute an inferior race incapable of self-government, that the American government was set up on a white man's basis, and that the unalienable rights of the Declaration of Independence are only unalienable for white men (see Johannsen, pp. 33, 45, 127–28, 196, 215–16, 299 [Douglas]). Still, we must ask ourselves how different their racial attitudes really were.

For the record, Douglas never denies that blacks are human beings, in some respects equal to whites. He actually prefers the same racial situation as Lincoln does: no slavery yet no equal citizenship for blacks. In fact, Lincoln, immediately after suggesting that Douglas is a racist, implicitly acknowledges their common humanitarianism. He invites Douglas to join the Republican Party "on principle"; that is, on the principle that slavery is wrong (see Johannsen, p. 21 [Lincoln]). Douglas, of course, never did. The reason, however, probably lay in his opposition to how the Republicans were presenting that principle to the public rather than in any doubts about the principle itself. Where Lincoln openly avows his conviction that slavery is wrong and that blacks are entitled to some, but not all, the same rights as whites, Douglas simply says that he approves of current Illinois law which prohibits blacks from being either citizens or slaves (see Johannsen, pp. 33, 46–47, 129, 266, 299 [Douglas]). He, then, does not deny humanitarian principles. What he denies is the presumption that his own (or Lincoln's or the Illinois majority's) humanitarian principles enjoy a privileged position in a liberal democracy that requires him to "interfere with" other people (or states) who hold different principles.

Nevertheless, it is possible that Douglas's rhetoric was not so much constrained by a particular understanding of liberalism or democracy as by a particular understanding of morality. Perhaps the significant difference lies not in the character of his racial attitudes but in their intensity. Lincoln's antislavery preferences do seem much more intense; much more central to his political ethic. The question is why.

Protestantism

Humanitarianism defines the wrong of slavery as the way it treats slaves merely as property. Both Lincoln and Douglas agreed to this claim, a claim which is inherently a moral one. Many moral claims, though, have multiple sources. It has been argued that Lincoln's moral claim against slavery also had a religious basis; that he also thought of slavery as a sin because it denied that slaves were men (and women) created in God's image; and that he, in particular, possessed a Protestant conscience which compelled him to bear witness to his moral claim against slavery. The implicit comparison is twofold: not only was Douglas's moral claim against slavery not similarly energized by a religious outlook, but a moral claim grounded strictly in humanitarianism is a psychologically weak one.

To be fair, the argument is not that Lincoln was a particularly religious person, any more than Douglas was. Neither, in fact, attended church.[55] The argument is rather that Lincoln was imbued with a powerful religious fatalism which was foreign to someone like Douglas, whose outlook was completely secular, and that Lincoln's largely untutored, yet identifiably Protestant, religious beliefs were at the core of his moral condemnations of slavery.[56]

There, again, is a certain truth to this argument. Especially as the crisis over slavery deepened and plunged the nation into civil war, a powerful religious fatalism did seem to move further and further to the forefront of Lincoln's public reflections on slavery. This tendency peaked in his haunting second inaugural. Even during his debates with Douglas, he condemned slavery in morally absolutist terms which appeared to betray a religious intensity lacking in his opponent. At Ottawa, Lincoln's long recitation from his Peoria speech expresses his hatred of slavery as a "monstrous injustice" and as teaching Americans that "there is no right principle of action but self-interest." At Galesburg, Lincoln quotes Jefferson on the subject of slavery to the effect that "'he trembled for his country when he remembered that God was just,'" language which, Lincoln correctly notes, Douglas would never use on the subject. Lincoln later denounces slavery as "a moral, social and political evil." Finally, at Alton, he characterizes "the real issue" of the debates as "the eternal struggle between these two principles—right and wrong—throughout the world. They are the two principles that have stood face to face from the beginning of time; and will ever continue to struggle" (Johannsen, pp. 50–51, 220, 226, 319 [Lincoln]).[57]

Without denying the increasing centrality of a nonsectarian religious fatalism to Lincoln's political ethic during the last years of his life, I would still contend that Protestantism played a relatively minor role in energizing his antislavery preferences. The evidence of specifically Protestant or religious thought is equivocal. Injustices are not monstrous and institutions are not morally evil in religious thought alone. Besides, Lincoln labels slavery a social and political evil as well as a moral one. Similarly, many categories of thought signal out "self-interest" for special suspicion. Lincoln, in the passages cited above, goes on to counterpose self-interest not to the Bible and its moral commandments but to the Declaration of Independence and its principles of civil liberty and to trace the eternal struggle between right and wrong not back to the Garden of Eden but to the divine right of kings. Indeed, the least equivocal evidence of religious thought in the Lincoln-Douglas debates is Lincoln's quote from Thomas Jefferson, a quote which is not even a good indication of the nature of Jefferson's political ethic.[58]

In short, the same evidence could be arrayed under several categories of secular thought, including republicanism and liberalism.[59] (I am even skeptical of the assumption that humanitarianism cannot account for this evidence.) Furthermore, the preponderance of evidence from the Lincoln-Douglas debates strongly suggests that both candidates thought of the slavery issue primarily (if not entirely) in secular terms. We must, then, look for the roots of the different intensities of their antislavery preferences in secular, not religious, thought. After referring to the biblical teaching on the universality of human selfishness, Lincoln himself quips that "I think we would have discovered that fact without the Bible" (Johannsen, p. 314 [Lincoln]).[60]

Republicanism

Perhaps republicanism played a larger role than humanitarianism and Protestantism in determining the intensity of Lincoln's antislavery preferences, both in themselves and in relation to Douglas's antislavery preferences. As we have seen, the republican revisionists argue that Lincoln thought slavery was wrong primarily because it elevated an ambitious Southern slavocracy to a position of inordinate economic and political power, thus destroying the relatively egalitarian society of America's republican past. In their view, the slavocracy defined the monstrous injustice of slavery for Lincoln, not the plight of the slaves, and the undesirable social consequences of the institution made it seem so evil to him, not its intrinsic immorality. Douglas differed from Lincoln in not harboring the same fears of a slavocracy. He, therefore, did not possess the same intense hatred of slavery and largely kept silent about the evils of the institution.[61]

I have already criticized this argument in the introduction of this chapter for slighting the moral side of Lincoln's politics. I can now add to that criticism. Although the revisionists rightfully point to Lincoln's concern with the aggregate consequences of slavery on white society, they misidentify the nature of the consequences with which he was most concerned. He did not appeal to a vision of a simpler republican past but to a vision of a future liberal society where all Americans participated in the benefits of equal liberty. This vision equally addressed the plight of the black slaves and the threat slavery posed to white Americans.

The republican interpretation of Lincoln can be strengthened. The revisionists actually overlook the aspects of his politics that are most congenial to their point of view, such as his conception of statesmanship. But, then, they still could not explain the depth of Lincoln's hatred of slavery, which was the grist of his own statesmanship. The republican interpretation cannot be strengthened enough to adequately account for the intensity of his antislavery preferences. It was not the specifically republican elements in Lincoln's political thought, nor the specifically pluralist ones, which explained the monstrous injustice of slavery to him. It was the generically liberal elements. For Lincoln, slavery was essentially an illiberal institution. It violated "the very fundamental principles of civil liberty," directly injuring its black victims and indirectly undermining the almost exemplary liberal society white men had established in America (Johannsen, pp. 50–51 [Lincoln]).

Liberalism

Both Lincoln and Douglas thought slavery was wrong primarily because it was an illiberal institution, but only Lincoln saw its continued existence as a "cancer" which attacked the vital spirit of a liberal society. This was the principal reason he condemned the institution in stronger terms than Douglas did. He considered slavery a unique threat to Americans' most fundamental political values, values which had, for the most part, been unarticulated in prior political debates.[62]

Slavery systematically deprived its black victims of both private and public liberty. Being good liberals, Lincoln and Douglas saw that deprivation as the primary injustice of slavery. Yet, they also realized that the maximum extension of liberty possible in any society is subject to "practical necessities." This realization informed Lincoln's understanding of the Declaration of Independence as a "standard maxim for free society" as it did Douglas's claim that black Americans should enjoy the basic rights mentioned in the Declaration "consistent with the safety [or good] of society" (cf. Johannsen, p. 304 [Lincoln]; pp. 33, 46, 216, 299 [Douglas]). Surprisingly or not, Lincoln and Douglas calculated the maximum extension of liberty possible in 1858 America and came to the same result. They approved of neither slavery nor citizenship for blacks. They, however, differed on the factor they considered the major road-block to black citizenship. For Douglas, it was racial inferiority; for Lincoln, it was color prejudice.[63]

Lincoln and Douglas, of course, were not the only Americans performing these calculations, and they also differed in how they reacted to that fact. Douglas believed that white majorities were, *prima facie*, justified in calculating that the safety of their society required racial slavery, even if, according to his own calculations, only the deprivation of black citizenship rights was socially necessary. Lincoln, conversely, believed that white majorities *might* be justified in depriving blacks of citizenship rights but that they never can be justified in depriving blacks of their basic private rights. Unlike Douglas, Lincoln refused to accept the latter deprivation as a matter of right, though he did accept it as a practical matter and on a temporary basis.

There are many reasons why Lincoln might have distinguished the ability of black Americans to enjoy private rights and citizenship rights in such a qualitative manner. One set of reasons assumes that citizenship rights are more important than private rights and, therefore, that the factors which might justify a white majority in depriving blacks of citizenship rights lose much, if not all, their force in the case of private rights. Lincoln might have thought racial inferiority prevented blacks from being good citizens and color prejudice prevented them from acting as a deliberative community in concert with white citizens. Similarly, he might have thought racial inferiority prevented blacks from being good shoemakers and color prejudice prevented them from selling their shoes at market rates to whites. Only the first state of affairs, however, seems to threaten the safety of society. And even if the first state of affairs justifies excluding blacks from the citizen body, it is hard to imagine how the second justifies enslaving them and foreclosing any chance of their becoming profitable members of society. As Lincoln saw it, the only possible justifications were illegitimate ones: the racism of white majorities and, what often comes to the same thing, their crude self-interest.[64] Douglas was much more willing to accept at face value the justifications other whites offered for enslaving blacks. Yet, it was Douglas, not Lincoln, to whom this set of reasons for distinguishing private and citizenship rights seemed more important.[65]

Lincoln, in contrast, emphasized a second set of reasons. This set assumes that some private rights are more important than citizenship rights and that those private rights are too important for whites *not* to share with blacks. Lincoln treated a limited number of private rights—the "unalienable" rights of the Declaration of Independence—as quintessentially liberal rights which (almost) everyone in a liberal society must possess for it to remain a liberal society. His emphasis on these basic liberal rights reveals not only how his principal objections to slavery were liberal ones but how he considered the institution a threat to both black and white Americans. The institution of racial slavery will, over time, deprive all Americans of the benefits of a liberal society by attacking the popular spirit of liberty upon which such a society rests. Thus, Lincoln, unlike Douglas, viewed slavery as a fundamental threat to a liberal America. However bad the denial of citizenship rights to blacks may have been, he felt it did not pose the same dangers to his society.[66]

What appeared to upset Lincoln the most about slavery was that it was a preliberal and hence an illiberal institution. After identifying the struggle between Douglas and himself with the eternal struggle between right and wrong, he provides a peculiarly modern (and secular) definition of that struggle. He asserts that it is a struggle between "the common right of humanity" and "the divine right of kings." His gloss on the latter principle is expansive.

> It is the same principle in whatever shape it develops itself. It is the same spirit that says, 'You work and toil and earn bread, and I'll eat it.' No matter in what shape it comes, whether from the mouth of a king who seeks to bestride the people of his own nation and live by the fruit of their labor, or from one race of men as an apology for enslaving another race, it is the same tyrannical principle. (Johannsen, p. 319 [Lincoln])

At bottom, Lincoln believes the institution of slavery wrongs the slaves by depriving them of control over their own lives. It blocks their "right to rise" through the fruits of their own labor.[67] Accordingly, the institution violates three of the fundamental tenets of liberalism: that every ("normal") man is rational enough to control his own life; that all men are created equal in that basic respect; and that each man should have the same opportunity to demonstrate any greater rationality he might possess and rise to his appropriate level in society. Douglas seems to deny these tenets in claiming that blacks are incapable of self-government, but he would never deny that they apply to white men.[68] It was preliberal thought and institutions that denied those tenets on a general basis. Lincoln's analogy of slavery to royal absolutism not only makes the wrong of slavery more concrete to white Americans, but it presents the continued existence of the institution as a real danger to them. After all, the "victory" of liberalism over royal absolutism must have seemed a very tenuous one to them in light of recent events in Europe and Latin America.

Lincoln first developed the analogy between slavery and royal absolutism in a campaign speech delivered in Chicago prior to the debates. Near the end of that speech, he argues that Douglas's public reading of the Declaration of Independence cribs the document to such an extent that it has no meaning for the city's growing immigrant population. Douglas would transform its universalistic message into merely a justification of Anglo-American independence. Lincoln also predicts that Douglas's reading of the Declaration (as well as of popular sovereignty and *Dred Scott*) would, if religiously followed, destroy the American system of government by undercutting the liberal assumptions of equal rationality and the popular sentiments of liberty upon which it depends.

> Now, I ask you in all soberness, if all these things, if indulged in, if ratified, if confirmed and endorsed, if taught to our children and repeated to them, do not tend to rub out the sentiment of liberty in the country, and to transform this government into a government of some other form. These arguments that are made, that the inferior race are to be treated with as much allowance as they are capable of enjoying; that as much is to be done for them as their condition will allow. What are these arguments? They are the arguments that kings have made for enslaving the people of all ages of the world. You will find that all the arguments in favor of kingcraft were of this class; they always bestrode the necks of the people, not that they wanted to do it, but because the people were better off being ridden.

Explicitly returning to Douglas's (mis)use of the Declaration, Lincoln suggests that its dangerous tendencies cannot be confined to blacks.

> I should like to know if taking this old Declaration of Independence, which declares that all men are equal upon principle, and making exceptions to it, where will it stop? If one man says it does not mean a negro, why may not another say it does not mean some other man?

Finally, in the peroration, Lincoln admonishes Americans to stop "all this quibbling about this man and the other man—this race and that race and the other race being inferior."[69]

As with most slippery-slope arguments, this one is problematic. We wonder how many white Americans felt personally threatened by the institution of slavery; how many feared that they, their children, or their children's children might be enslaved, that slavery would one day be legal in all the states, or that their government was in imminent danger of becoming an absolute monarchy.[70] Admittedly, Lincoln's lesson is somewhat more subtle. The immediate danger is that racial slavery will be accepted "upon principle." That acceptance, though, will already signal a major slippage in popular sentiments of liberty. As that slippage continues, the nation will eventually be unable to sustain a liberal government. The loss of the blessings of that form of government, and the evils of another form of government, defines the principal threat of slavery to white Americans. In the most ominous of his many metaphors for slavery, Lincoln likens it to a cancer. It is the only institution that "has ever threatened our liberty and prosperity" (Johannsen, p. 317 [Lincoln]).

Lincoln not only accuses Douglas of helping the cancer spread physically throughout the body politic but, even more, of weakening its spiritual resources to combat the disease. Douglas's "I don't care" attitude toward the spread of slavery and his cribbed reading of the Declaration of Independence are undermining Americans' spirit of liberty (see Johannsen, pp. 66–67, 233–34 [Lincoln]). Lincoln seizes the opposite role. He is struggling to contain the cancer, both physically and spiritually (since, to continue the metaphor, there is no known cure which would not also be fatal to the body politic). His free-soil policies and his sweeping reading of the Declaration are primarily directed to preventing any slippage in Americans' spirit of liberty. Indeed, all his major doctrines appear to be primarily directed to preventing that possibility, a possibility which would be truly tragic, obviously for the slaves but equally, or so Lincoln insists, for white Americans.[71]

Yet, we still wonder how many white Americans found *this* slippery slope credible. How many feared that their liberal society would be the ultimate victim of the institution of slavery? Douglas certainly did not. He thought that whatever evils festered within the institution, they did not threaten white Americans. He was convinced that racial slavery, notwithstanding its illiberal character, could exist indefinitely within a liberal society without putting that society at risk.[72] While Lincoln agreed that slavery would continue to exist indefinitely within his own liberal society, he did not think its evils could be so easily contained. He claimed that the minimum requirement for his society to retain its liberal character was a renewed commitment to the belief that slavery is wrong, and a wrong which must eventually be expunged from American soil.

It is difficult to say whether Lincoln or Douglas better gauged the future, since Southern secession, civil war, and enforced emancipation intervened. However, it is clear that their different political forecasts were influenced by their different understandings of the threat slavery did, or did not, pose to white liberal America. Only Lincoln saw a liberal society as requiring equal liberty and only he saw the institution of slavery as representing a massive violation of equal liberty. In stark opposition to Douglas, Lincoln used the Declaration of Independence to promulgate the idea that equal liberty is a collective good that can only exist in a society whose members equally enjoy certain basic rights.[73] Racial slavery not only deprives individual blacks of their basic rights, but it deprives a whole society of equal liberty. Lincoln also used the Declaration to publicly acknowledge that a gap will always exist between this ideal and the real conditions of a liberal society. In a truly liberal society, though, that gap must somehow be manageable, it must be shrinking, not growing, and a solid majority of the citizens must be committed to further closing it.

Liberalism, then, lay at the root of Lincoln's antislavery preferences. It provides an explanation of his position on the slavery issue which combines the truths of the three previous explanations. In distinction to explanations based on humanitarianism, Protestantism, and republicanism, it integrates Lincoln's moral attacks on slavery with his political ones. Although his position was not simply a liberal one, it was more of a liberal position than it was a humanitarian, Protestant, or (specifically) republican position.[74] Liberalism is the factor which best explains the character and intensity of his antislavery preferences and why he condemned the institution in stronger terms than Douglas did.

Nonetheless, we need to consider yet another factor. Whereas Lincoln's and Douglas's different understandings of liberalism is the primary factor in explaining how their preferences on the slavery issue differed from each other, their different understandings of democracy is the primary factor in explaining how their principles of action on the issue differed from each other. No matter how many different ways we look at their preferences on the issue, those preferences appear relatively similar. It is their principles of action which appear conflictual. Lincoln did not merely condemn slavery in stronger terms than Douglas did. Douglas did not publicly condemn slavery in any terms and he counseled against any one else doing so on the grounds that it was undemocratic.

Democratic Statesmanship

Republicanism and pluralism, which were not very helpful in trying to explain the character and intensity of Lincoln's antislavery preferences, will be very helpful in trying to explain his understanding of democracy and how it differed from Douglas's. Both of their understandings were essentially pluralistic, but insofar as they differed—which they did to a significant extent—they differed because Lincoln's conception of statesmanship more strongly reflected traditional republican views of statesmanship. Just as Douglas's critique of Lincoln's statesmanship became one of the major motifs of his campaign, Lincoln deprecated Douglas's statesmanship, or, according to Lincoln, lack of statesmanship. He, however, first had to parry Douglas's critique.

As we saw in chapter 7, Douglas believes interstate and interpersonal comity are fundamental democratic values. He charges Lincoln with violating those values by making war on the institutions of other states and by failing to maintain a statesmanlike public indifference toward the fate of slavery in America. Lincoln naturally denies the charge. He portrays himself and his party as practicing considerable self-restraint on the slavery issue. In part, this self-restraint is simply a matter of constitutional theory. Like Douglas and most other Americans at the time, Lincoln believes the Constitution severely limits how states and citizens can act toward one another, as well as how the federal government can act on the state governments and individual citizens. In his own mind, his party's free-soil policy is a testament to this constitutionally mandated interstate and interpersonal comity. It leaves the institution of slavery alone in the states where it already exists and with the people who are currently enjoying its forbidden fruits (see Johannsen, pp. 131–32, 221, 254–55, 315 [Lincoln]).[75]

Interstate and interpersonal comity, though, is not just a matter of constitutional theory for Douglas. It is also a matter of democratic theory. Those two values call attention to the respect any democratic statesman should show toward the opinions of his audience, within and across state boundaries. Lincoln must indicate how he honors interstate and interpersonal comity in this less literal, extraconstitutional sense.

Beyond renouncing all intentions to legally interfere with slavery in the Southern states, Lincoln can claim to have always been solicitous of the preferences and interests which support the institution in those states. Once more, his recitation of his Peoria speech offers crucial evidence.

> Before proceeding, let me say I think I have no prejudice against the Southern people. They are just what we would be in their situation. If slavery did not now exist among them, they would not introduce it. If it did now exist amongst us, we should not instantly give it up . . . When Southern people tell us they are no more responsible for the origin of slavery than we, I acknowledge that fact. When it is said that the institution exists, and that it is very difficult to get rid of it, in any satisfactory way, I can understand and appreciate the saying. I surely will not blame them for not doing what I should not know how to do myself. (Johannsen, p. 51 [Lincoln])[76]

Yet, as a matter of democratic theory, interstate comity largely collapses into interpersonal comity.[77] Douglas is not just concerned with being solicitous to Southern opinion. He is also concerned with being solicitous to Northern opinion and to Illinois opinion. The Peoria quote suggests that Lincoln is equally concerned. Thus he posits the impossibility of a biracial society in America by saying that "[m]y own feelings will not admit of this; and if mine would, we well know that those of the great mass of white people will not." After shifting the burden from his own preferences to his audience's, he adds: "Whether this feeling accords with justice and sound judgment, is not the sole question, if, indeed, it is any part of it. A universal feeling, whether well or ill-founded, cannot be safely disregarded." Later in the same debate (no longer quoting from his Peoria speech), he reinforces this message. "In this and like communities, public sentiment is everything. With public sentiment, nothing can fail; without it nothing can succeed" (Johannsen, pp. 51, 64–65 [Lincoln]).

Except for the more philosophic language, Lincoln sounds very much like Douglas in his solicitousness toward public opinion. He presents himself as a democratic statesman who is severely limited by the nature of public sentiments. He views the most formidable obstacle to translating his own antislavery preferences into law as the divergent preferences of other Americans on the slavery issue, not any guarantees for states' rights or property rights which may exist in the Constitution. He, in effect, grants the same precedence as Douglas does to the public liberty of white majorities to determine what rights, if any, blacks shall enjoy over the private liberty of blacks to enjoy certain unalienable rights independent of the determinations of white majorities.

Lincoln, nevertheless, believes that, in principle, the private liberty of blacks takes precedence over the public liberty of whites, a belief which, I have argued, is primarily rooted in his liberalism. This belief places him in direct opposition to Douglas. His formulation of that opposition is justly famous. "He [Douglas] contends that whatever community wants slaves has a right to have them. So they have if it is not a wrong. But if it is a wrong, he cannot say people have a right to do wrong" (Johannsen, p. 319 [Lincoln]).[78]

Given the tension Lincoln identifies between the practical priority of the public liberty of popular majorities in liberal democracies and the ethical priority of the private liberty of individual citizens (or noncitizens) in such societies, he is confronted with the problem of ensuring that at least the most basic private rights take precedence over majority rule. For Lincoln, however, it is more a question of concurrence than of precedence. He thinks "it is extremely important" both that the people "shall decide" and that they shall "rightly decide" (Johannsen, pp. 236–37 [Lincoln]). He seems very confident that this concurrence of majority preferences on the right decision can be achieved through persuasive, democratic means; that is, without violating interstate and interpersonal comity. In part, his confidence is based on his belief in an antislavery consensus in America so that it is less a question of changing preferences than of bringing latent preferences to the surface. Still, Douglas's claim that Lincoln's rhetoric violates interstate and interpersonal comity contains a large measure of truth.

In a campaign speech delivered prior to the debates, Douglas phrased his critique of Lincoln's statesmanship in a particularly felicitous way. As repeated by Lincoln in a subsequent speech, Douglas charged "that while I had protested against entering into the slave States, I nevertheless did mean to go on the banks of the Ohio and throw missiles into Kentucky, to disturb them in their domestic institutions."[79] Lincoln, of course, only repeats the charge to deny it. Many of his public statements on the slavery issue, though, did amount to throwing (verbal) missiles into Kentucky and the other slave states. The Peoria quote contains implicit criticisms of Southerners for holding slaves in the first place and then for not establishing programs of gradual emancipation (see Johannsen, pp. 51–52 [Lincoln]). His party's free-soil policy may not have legally interfered with the institution of slavery in the Southern states but rhetorically it was a time-bomb. The stated purpose of the policy is to treat slavery as a disease. The implication is that the slaveholders are vermin. Perhaps Lincoln was surprised at how strongly many Southerners reacted to his rhetoric in view of the moderate nature of his language and immediate policy goals. There, however, was really nothing surprising about their reaction. To them, his rhetoric was a clear violation of interstate and interpersonal comity. It questioned the legitimacy of their institutions, their interests, and their morals. As Douglas noted, the Republican Party could not promulgate its doctrines in the South. Whatever we may think about Lincoln's response that public opposition is hardly sufficient proof of the falsity of a doctrine, the fact remains that his party had no measurable support south of the Mason-Dixon line (see Johannsen, pp. 222–23 [Lincoln]).[80] To Douglas, that fact proves the party cannot stand the test of a national democracy. Lincoln's response also must have struck him as suspiciously undemocratic because it presumes that the Republican Party is the vehicle of a higher truth the people at large are too blind, or too corrupt, to see.

It is open to Lincoln to argue that he and his party are not violating generally accepted notions of interstate and interpersonal comity. After all, he is just talking about slavery and nothing he is saying appears to trespass on the norms of democratic discourse. He, moreover, is not imposing his antislavery preferences on other citizens. They are perfectly free to reject (or accept) his public statements on the slavery issue. Lincoln does make these points. But he also takes a more positive tack. He argues that Douglas's democratic intuition is fundamentally flawed. It precludes the genuine dialogue necessary to even a minimally functioning democracy. What Douglas sees as a violation of interstate and interpersonal comity, Lincoln sees as the essence of democratic statesmanship; what Douglas sees as a statesmanlike self-restraint, Lincoln sees as a missed opportunity. According to Lincoln, it not only is true that a statesman cannot fail with the support of public sentiment. It also is true that "he who moulds public sentiment, goes deeper than he who enacts statutes or pronounces decisions. He makes statutes and decisions possible or impossible to be executed" (Johannsen, p. 65 [Lincoln]).

Lincoln claims that Douglas's "statesmanship" is really a lack of statesmanship. Lincoln develops this critique in two directions. First, he contends that Douglas is not using his great popular influence for good purposes. Second, he contends that Douglas is, in fact, using his great popular influence for bad purposes.

For Lincoln, Douglas clearly is someone who could make the execution of "statutes and decisions possible or impossible." His failure to do so can only be understood as a lack of statesmanship. Lincoln develops this critique in the course of excoriating Douglas's public stance on the *Dred Scott* decision.

> . . . Judge Douglas is a man of vast influence, so great that it is enough for many men to profess to believe anything, when they once find out that Judge Douglas professes to believe it. Consider also the attitude he occupies at the head of a large party—a party which he claims has a majority of all the voters in the country. This man sticks to a decision which forbids the people of a Territory from excluding slavery, and he does so not because he says it is right in itself—he does not give any opinion on that—but because it has been decided by the court, and being decided by court, he is, and you are bound to take it in your political action as law—not that he judges at all of its merits, but because a decision of the court is to him a 'Thus saith the Lord.' (Johannsen, p. 65 [Lincoln])[81]

The pattern is similar on the broader issues involved in the escalating party and sectional dispute over slavery. Lincoln observes that the Democrats' attitude of public indifference toward the expansion and indefinite perpetuation of the institution excludes treating it as a wrong, referring to it as a wrong, and, ultimately, even talking about it at all (see Johannsen, pp. 225–26, 255–57, 317–19 [Lincoln]). He suggests that many Democrats—probably alluding to Douglas himself—think slavery is wrong but that they are misplaced in a party that has no platform upon which to say it is wrong (see Johannsen, pp. 256, 318 [Lincoln]). Douglas and other Democratic leaders seem to assume that the crisis will pass if only Americans stop talking about slavery. Lincoln demurs. As long as the institution exists, it will be an issue because it grates on "the moral constitution of men's minds" (Johannsen, p. 54 [Lincoln]). All this leads Lincoln to, again, impugn Douglas's lack of statesmanship.

> But where is the philosophy or statesmanship which assumes that you can quiet that disturbing element in our society which has disturbed us for more than half a century, which has been the only serious danger that has threatened our institutions—I say, where is the philosophy or the statesmanship based on the assumption that we are to quit talking about it, and that the public mind is all at once to cease being agitated by it? Yet, that is the policy here in the north that Douglas is advocating—that we are to care nothing about it! I ask you if it is not a false philosophy? Is it not a false statesmanship that undertakes to build up a system of policy upon the basis of caring nothing about the very thing that every body does care the most about?—a thing which all experience has shown we care a very great deal about? (Johannsen, p. 315 [Lincoln])

Lincoln, though, accuses Douglas of more than merely a lack of statesmanship. His second critique is that Douglas is using his great popular influence to help perpetuate the institution of slavery. Lincoln claims that logically Douglas and other Democrats cannot say they are indifferent to the fate of slavery if they think slavery is wrong because no one can be indifferent to a wrong. Similarly, they cannot logically say they are as indifferent to the spread of slavery into the new territories as they are to the spread of other forms of property unless they think slaves are just like other forms of property. Lincoln insists that it does not matter to him whether they actually hold those beliefs or not because the effect is the same. Their rhetoric is preparing the public mind to think slavery is not wrong and, thus, to accept the institution as a permanent feature of the American landscape (see Johannsen, pp. 225, 256–57, 319 [Lincoln]).

Douglas is the target of special criticism. At the end of the "House Divided" speech, Lincoln refers to the incipient movement among Republicans to run Douglas as a fusion candidate, in part because his national prominence would allegedly enable him, better than a senator Lincoln, to block the proslavery policies of the Buchanan administration. Lincoln denigrates Douglas's ability and desire to become such an antislavery instrument, asserting that so far he "has done all in his power to reduce the whole question of slavery to one of a mere right of property." Later in the debates, Lincoln calls Douglas "the best instrument" for "preparing (whether purposely or not) the way for making the institution of slavery national." During the final debate, Lincoln, then, declares that "willingly or unwillingly, or purposely or without purpose, Judge Douglas has been the most prominent instrument in . . . putting it [slavery] upon [South Carolina Rep. Preston] Brooks's cotton-gin basis—placing it where he openly confesses he has no desire there shall ever be an end of it." Fundamentally, however, Lincoln denounces Douglas for undercutting the spirit of liberty upon which the future end of slavery depends. Through his public reading of the Declaration of Independence, "he is penetrating, so far as lies in his power, the human soul, and eradicating the light of reason and the love of liberty . . . [and] in every possible way preparing the public mind, by his vast influence, for making the institution of slavery perpetual and national" (Johannsen, pp. 20, 233, 320, 233–34 [Lincoln]).[82]

Recalling that Lincoln believes that public sentiment is everything in a democracy and that the particular public sentiment of liberty is everything in a liberal democracy, Douglas plainly is the villain of the piece. Yet, his villainy is ambiguous. Upon closer inspection, Lincoln's second critique of Douglas's statesmanship turns out to be a more sophisticated version of his first critique. It is the same lack of statesmanship which explains why Douglas is using his great popular influence for bad purposes that explains why he is not using it for good purposes.

The second critique goes to the heart of Lincoln's conspiracy theory. Lincoln identifies a certain coincidence between Douglas's actions and the actions of those who do desire to perpetuate the institution of slavery. This coincidence need not be conspiratorial in nature; nor is Douglas necessarily a pawn of the proslavery forces. Although Lincoln wavers on making those charges, the thrust of this critique is that Douglas (and perhaps equally the proslavery forces) seems unaware of the long-term consequences of his actions. The charge is, again, a lack of statesmanship. Lincoln rebukes Douglas for his lack of foresight, not for his evil intentions. Douglas has consistently failed to take a broad enough perspective on his actions to see how they might be contributing to an outcome he himself would find unacceptable.[83]

Lincoln opens the "House Divided" speech on a speculative note, a note which cues his audience to how it should interpret the conspiracy theory he proceeds to articulate. "Mr. President, and Gentlemen of the Convention: If we could first know where we are, and whither we are tending, we could better judge what to do, and how to do it" (Johannsen, p. 14 [Lincoln]). To engage in such public speculation seems to be one of the first duties of statesmanship. During the course of the debates, Lincoln continually questions Douglas's disinclination to fulfill this duty. Lincoln, thus, asks Douglas and his political allies if in denying the Declaration of Independence applies to black Americans they are "not being bourne along by an irresistible current—whither, they know not?" (Johannsen, p. 305 [Lincoln]).

Now, it is doubtful that Douglas was especially blind to the consequences of his actions. In the first place, he did not share Lincoln's assessment of what those consequences were likely to be. He, on the contrary, thought his actions were more likely to produce the outcome both he and Lincoln desired than Lincoln's actions were. He did studiously avoid making public statements to that effect, but it was not for the reasons Lincoln attributed to him. In the second place, then, Douglas considered the "whither are we tending" question strongly undemocratic insofar as the implication was that some citizens had privileged access to the right answer and that those citizens should be granted the political authority to lead the citizen body in the indicated direction. He was not really insensitive to the question, nor was he really indifferent to the answer. He simply believed that it was the citizen body's right to answer the question for itself. His popular-sovereignty doctrine would enable it to do so. If, *ex hypothesi,* his actions were promoting "slavery national," it was because his actions were allowing the citizen body to choose "slavery national," not because he himself wanted it to move in that direction or because he was inadvertently pushing it in that direction. In short, Douglas thought the usual reasons for asking and acting on the "whither are we tending" question were elitist.[84]

While Douglas believed Lincoln's concerted efforts to prejudice "freedom national" were undemocratic, Lincoln believed those efforts were positively enjoined by his role as a democratic statesman. He was "merely" attempting to guide public opinion in the right direction. The different ways he and Douglas reacted to the *Dred Scott* decision substantiate these different understandings of democracy and democratic statesmanship.

Douglas does treat the decision as a "thus saith the Lord." The Court has spoken. Actually, the American people through their representatives on the Court have spoken. There is no need for any more public dialogue on the case. American citizens who disagree with the decision have two options. They can either accept the decision and adjust their political views accordingly or they can disobey the decision and subvert the democratic processes of government through mob rule (see Johannsen, pp. 160–61, 242–44, 267–69 [Douglas]).[85] Lincoln claims that Douglas's definition of the options is too stark. Without disobeying the decision, citizens can still criticize the decision and try to overturn it (see Johannsen, p. 255 [Lincoln]).[86] This reaction is not an appeal to mob rule but part of a continuing democratic dialogue over the issues involved in the case. Lincoln naturally feels that it is incumbent upon himself not to accept the finality of the *Dred Scott* decision because it is antithetical to his party's free-soil policies. The decision, however, also seems inconsistent with Douglas's popular-sovereignty doctrine. Lincoln queries Douglas as to why he, too, does not expressly reject the decision as a political rule since under it "squatter sovereignty squatted out of existence" (Johannsen, p. 16 [Lincoln]).[87] His failure to do so indicates just how much his and Lincoln's understandings of democracy differ.

Lincoln argues that no one in a political community can be indifferent to whether it makes the right or wrong decision on an issue of such immense proportions as the slavery issue. Everyone in the community, therefore, has an obligation to try to ensure that it makes the right decision. If someone occupies a position of political influence, then all the more should he act in that manner so long as he does not violate the basic norms of a democratic society. To Lincoln, Douglas is perversely denying his own political role through an exaggerated understanding of those norms.

It seems impossible not to be more sympathetic to Lincoln's critique of Douglas's statesmanship than to Douglas's critique of Lincoln's statesmanship. The tendency is to caricature Douglas's position. Three factors, though, speak in its favor. First, his position appears more persuasive to the extent that we consider the beliefs of antebellum Americans about the rightness or wrongness of slavery merely personal preferences. Douglas certainly seems determined to reduce the one to the other, thereby deflating the whole issue. Second, his position appears more persuasive to the extent that we are convinced American public opinion would have automatically moved in an antislavery direction; that is, independent of any organized efforts to guide it in that direction. Douglas clearly thinks that public opinion is more responsive to such material factors as economic profit than it is to public philosophies or government policies and that the one will eventually condemn slavery more surely than the other two. Third, his position appears more persuasive to the extent that we accept his implicit analogy between political democracy and a free-market economy. Douglas not only believes that an "automatic" process of opinion change is the most democratic process but that it will produce the best results. The role of the democratic statesman is to protect the integrity of such a process, not to attempt to control it. This intuition is the ultimate measure of Douglas's pluralism. It taps a strong strain in American democratic thought which defines the alternative, teleological model of the democratic process as elitist.

We, nevertheless, are more sympathetic to Lincoln's position precisely because it taps both of the major strains in American democratic thought. His position reflects both the essentially pluralistic, "free market" model of the democratic process and the essentially republican, teleological model. He rejects, or, rather, only accepts as partial truths, the three key assumptions of Douglas's position. Lincoln does not feel that prevailing beliefs about the wrongness (and also rightness?) of slavery are merely personal preferences; that public opinion will automatically move in an antislavery direction; or that political democracy is analogous to a free-market economy. Again, a specific case will help reveal how his understanding of democracy was less pluralistic than Douglas's understanding.

This case involves the contrasting ways they foresee slavery being abolished in America. Douglas assumes that the institution will not expand into the new territories and that it will slowly die a natural death in the South because it will become increasingly unprofitable. He, thus, ascribes Illinois's abolition of slavery to economics, not to the antislavery predispositions of its early settlers or to the legal (and moral) force of the Northwest Ordinance (see Johannsen, p. 299 [Douglas]). Lincoln, conversely, assumes that the Northwest Ordinance was critical to his state's rejection of slavery, just as a legally enforced free-soil policy will be critical to the rejection of slavery in the states carved out of the territories further west (see Johannsen, pp. 77–78, 316 [Lincoln]).[88] As compared to Douglas, Lincoln seems less optimistic that in the competition between free and slave labor, free labor will always emerge victorious. He attacks Douglas's "Freeport Doctrine" on these grounds. The doctrine presumes that, to thrive, slavery requires special legal protection. Yet, the history of the country "shows that there is vigor enough in slavery to plant itself in a new country even against unfriendly legislation. It takes not only law but the enforcement of law to keep it out" (Johannsen, p. 147 [Lincoln]). The crucial factor for Lincoln, however, is not the law or even the enforcement of the law but the public sentiment which makes those things possible.

The case, then, does not so much involve different expectations about the future as it does different understandings of democracy. According to Lincoln, the democratic process is not *automatically* progressive. Citizens and statesmen must consciously act to make it so, though he does seem generally confident that they will act in the appropriate manner.

By the same token that Lincoln's understanding of democracy is less pluralistic than Douglas's understanding, it is more republican. Lincoln appears uncomfortable with Douglas's "free market" model of democracy, favoring instead the older model of democracy as a consciously directed, goal-oriented process which prescribes positive (not just negative) duties for citizens and statesmen. In Lincoln's understanding of democracy, we find strong analogues to traditional republican conceptions of secular history, civic virtue, political community, natural aristocracy, and the public good. Where Douglas only sees progress, Lincoln also sees the possibility of regress. Where Douglas insists "that if any one man choose to enslave another, no third men shall be allowed to object," Lincoln insists that the third man not only has a duty to object but he has a stake in what the first man does to the second because they are all part of the same community (see Johannsen, pp. 15, 315 [Lincoln]). And where Douglas cautions statesmen against imposing their preferences on other equal citizens, Lincoln urges them to direct public opinion toward the goal of equal liberty.

In his republicanism, Lincoln was a political anachronism. He was the last of the great republican statesmen in America. The future belonged to the Douglases and ever more extreme forms of pluralist democracy. Still, the Lincoln myth would not continue to resonate today if he was simply an anachronism. His understanding of democracy was more pluralistic than republican and more pluralistic than the understandings of most of his predecessors.[89]

Lincoln's paramount concern with public opinion distinguished him from his more republican predecessors. Unlike the Federalists and the Websters; the Calhouns and the Anti-Federalists, Lincoln assumed a close, symbiotic relationship between citizens and statesmen. He dismissed the traditional deference of citizens to statesmen which had long been thought necessary to a well-ordered democracy. If anything, statesmen must defer to public opinion, although Lincoln did not embrace the flip side of this pluralistic development nearly as much as Douglas did. Both Lincoln and Douglas, moreover, were men of the people, exemplars of the new type of American politician.[90]

Lincoln did not repudiate Douglas's "free market" democracy so much as hesitate at its more extreme implications. His appeals to civic virtue were certainly palpable on the slavery issue, yet in terms of concrete actions they seemed to dissolve into a general obligation to maintain a spirit of liberty. Similarly, Lincoln's theory of nationalism mostly yielded to the reality of a pluralist society. The substantive vision of the public good which he did not yield to that reality, again, only appeared substantive on the slavery issue—an issue which, he believed, uniquely shook the liberal foundations of his society. Finally, notwithstanding all his forebodings about the future, Lincoln's idea of history remained essentially progressive. He was confident that once the issue was clearly drawn between slavery and freedom, the antislavery forces would prevail.

In the end, we must be impressed with the complexity of Lincoln's antislavery politics. His position on the slavery issue combined elements not only of pluralism and republicanism but of nationalism, consensus social realism, filiopiety, history, humanitarianism, Protestantism, generic-liberalism, and democracy. His basic approach to politics was not as intuitive as Douglas's; nor was it as historical as Webster's, as metaphysical as Calhoun's, as bold as Publius's, or as cautious as the Federal Farmer's. Once again, he seemed to synthesize all those dispositions into a greater whole. The complexity of his politics accounts for his remarkable political career. It also constantly renews the Lincoln myth.[91]

9

Republicanism as Bad Conscience

The end of the Civil War and Lincoln's tragic death marked the watershed between American republicanism and pluralism. Yet, republican themes persisted in American political thought, and the nation's republican past continued to provide Americans with a critical perspective on their pluralistic politics. Republican "revivals," moreover, occurred during the Progressive era and the 1960s. In this brief, concluding chapter, I will outline this postbellum history.

The Civil War inaugurated, or, more, consummated, dramatic changes in American politics. It is not difficult to account for this fact within my framework of analysis. The war provided a tremendous impetus to the two factors I have identified as crucial to the shift from predominantly republican to predominantly pluralist modes of thought. It meant, literally and emotionally, the triumph of American nationalism. For the next hundred years, federalism and states' rights were strongly associated with racism in the minds of many Americans. Nationalist statesmen felt less urgency to magnify the union, thus transcending what had previously been the central debate within American republicanism.[1] The war also unleashed the powerful pluralistic forces which had been latent within American society, both through the traditional social structures it destroyed and through the new ones it created. On a more mundane level, it witnessed the death of a political generation, quickening the pace of such evolutionary processes as the shift from republicanism to pluralism within the American liberal tradition. Finally, to the survivors, who had participated in the massive arming of civic virtue at the beginning of the war, its end could only usher in a period of disillusionment. There was a characteristic turning from public to private concerns, a turning which was characteristically pluralistic.[2]

The Progressive era actually saw three interrelated developments.[3] First, the Progressive reformers repudiated what they viewed as the increasingly corrupt politics of the post–Civil War years. They sought a return to the principles of the original republic, although they, of course, did not reject "good" progress. They, then, did not simply oppose a republican past to a pluralist present. They, however, did largely define reform as revival, and what they wished to revive clearly evoked republican themes. Predictably, they enjoyed their greatest successes on the local level, where a republican politics still had some concrete meaning in identifiable public goods, in easily perceived nexuses between private sacrifices and the attainment of those goods, and in civic leaders strongly committed to that process. Second, the academic side of this reform movement rewrote American history as a morality play. The Progressive historians interpreted the past as a battleground between alternately victorious democratic and nondemocratic forces. Again, their work can, to a considerable extent, be recast in terms of the opposition between a republican and a pluralist politics. Third, and conversely, the nascent social sciences were prepared to unequivocally embrace a pluralist politics. Insofar as they showed any interest in history, the early pluralists interpreted the past in terms of the present. For the most part, though, they abandoned the past for a more pluralistic, and democratic, future. It soon became *passé* for those engaged in the *scientific* study of politics to refer to such undefinable concepts as the public good and civic virtue, especially after the hands-on experience in "big" government many of them received during World War I. Instead, such "operational" concepts as interest groups and group interests, responsive elites and latent publics came to dominate their political discourse. Those actively engaged in the practice of politics were not broken of old habits as quickly or completely. They merely used the newer pluralist rhetoric alongside the older republican rhetoric.

For both political scientists and politicians, the 1950s was the heyday of pluralism. Truman (David not Harry) updated Bentley, and Ike recalled the pubescent pluralism of the Republican administrations of the 1920s.[4] The interpenetration of government and academia which mushroomed during World War II, once again, had a significant impact in strengthening the grip of pluralism on both institutions. In the next decade, however, a reaction occurred, and the end of pluralism (liberalism?) was confidently proclaimed.[5] As part of a highly vocalized search for political community—a search which has even made inroads on the scientific study of politics—Americans today are still struggling with the legacies of the 1960s. The ecology movement, volunteerism, and local politics (including community participation in state and federal programs) all have been celebrated as new, or renewed, settings for the practice of civic virtue.[6]

In assessing the nature of these developments, it is important to recognize that America is a liberal society and that it has been one ever since it first took shape as a distinctive society (or societies) in the early 1700s. It is a society in which statesmen have always felt uncomfortable telling their audiences what to do. How can they, in one breath, say this is a free country and, in the next, tell us how to use our freedom? More than any other American statesmen, Lincoln epitomized this dilemma; Douglas evaded it. To assume that America once was a drastically different type of society not only is historically inaccurate but is bound to distort our evaluation of current trends. The search for political community in contemporary America is not the search for a neoclassical *polis*. It is the search for a liberal community. Few Americans today, certainly not presidents, but even not most academic communitarians, want to give up liberalism. What they want is a kinder, gentler liberalism.[7]

While an integral part of my argument is that America once was a different type of liberal society in which statesmen were more likely to attempt to guide, rather than just react to, public opinion, the counterpoint remains important. The processes underlying the pluralistic transformation of American politics have not made the search for a kinder, gentler liberalism simply quixotic, but they have made it increasingly difficult. American statesmen can still draw on the republican past. That past, though, does seem less and less usable; more and more anachronistic.

We can substantiate the anachronistic nature of the republican past by returning to our discussion of the sixteen core-beliefs of republicanism (see chapter 2). As good liberals, Americans are still champions of liberty and happiness. However, the liberty and happiness they champion seldom have public connotations, so much so that "public liberty" has become almost an archaic usage. The most conspicuous outlet of civic virtue today is the "Not In My Backyard" syndrome, a syndrome which is very paradoxical from a republican perspective because its participatory thrust seems so narrowly self-interested. Fears of the decline of the American republic are probably as prominent now as they have ever been. Those fears, though, mostly take an economic form, as in our alleged inability to compete with Japan in world markets. Luxury also receives its share of criticism, but that criticism is, again, usually framed in terms of economic equity and fairness, not in terms of moral decay. Suspicion of government remains strong, yet it does not appear to be counterbalanced by any appreciation of the elevating potential of politics. While the melting pot vies with cultural diversity, the demand for social homogeneity is truly a reactionary one. Commerce is only problematic because of our shortcomings, not our successes. Small may be beautiful as an ideal. In an increasingly Balkanized world, it becomes a source of anxiety. The political-party system is frequently accused of being too weak, rarely of being too powerful. Today, democracy means that government representatives are errand boys (and girls). The republican vision of representation as a mixture of judgment and instruction has become more and more one-sided. This one-sidedness also characterizes prevailing beliefs about political corruption and republican jealousy. The predominant sentiment now is to hate Congress *and* our congressmen, although we still keep voting for them because they deliver the goods.[8] We hear echoes of republicanism in current political controversies. But we no longer hear full-blown republican symphonies.

What, for example, is the character of local politics? A question of scale obviously arises in the practice of civic virtue. In the late eighteenth and early nineteenth centuries, such American statesmen as the Federal Farmer and Calhoun bitterly resisted political consolidation on behalf of the ideal of a small-scale republican politics. Republican revivalists now place their hopes in the reinvigoration of that ideal.[9] Yet, it is debatable whether local politics was ever more elevated than, or even essentially different from, national politics. Publius and Webster, after all, argued that it was far worse.

In the late twentieth century, local politics is national politics writ small. Instead of the United Auto Workers lobbying Congress for restrictions on foreign-car imports, the Woodlawn Organization lobbies city hall for more police protection. In a sense, it has always been this way. Nevertheless, significant, even if subtle, changes in political practice and rhetoric—or, perhaps, merely in the public spirit or philosophy behind the very same activities and words—have taken place in response to broader social changes. Few cities or towns remain socially homogeneous and it is more difficult to provide any meaning to claims (whether disingenuous or not) that certain policies benefit the whole community. Local groups are no longer reticent about demanding shares of community resources on the basis of numbers, or need, or just plain clout. Progressive reformers labored to establish at-large elections to discourage the pluralistic politics such demands engendered. Racial and ethnic groups now condemn at-large elections for dissipating their political power. Despite asides to local responsibility, the "new federalism" has largely been debated in terms of "who pays," not "who decides."[10] Everyone knows who decides! However we evaluate this state of affairs, it is hardly a strong incentive to vote in state and municipal elections. The overarching point is that participation rates in local politics remain disappointingly low.[11]

It would be relatively easy to debunk the volunteerism and ecology movements in similar terms. My aim, though, is not to debunk any of these developments, nor is it to argue that they contain no potential to reinvigorate community politics or to insist that they are only pale imitations of their eighteenth- and nineteenth-century republican precedents. From a sufficiently enlarged point of view, they are not really developments at all. They are merely the most recent eruptions of political virtue upon an intractable core of economic interest. What I find curious is not the consistency of the mix of virtue and interest in American politics but rather the different strategies American statesmen have historically adopted to try to refine that mix. Preliberal efforts to cultivate an ascetic disinterestedness soon gave way to a variety of identifiably liberal-republican efforts that simultaneously appealed to virtue and interest. These approaches, in turn, were replaced by more pluralistic ones, which essentially provided an imprimatur to interest until such time that any distinction between interest and virtue was denied. It seems natural that the cycle would then reverse itself. Yet, as long as America remains a liberal society, the pendulum will never swing back very far.

In this study, I have sought to elucidate why the ultimately moral strategies of American statesmen evolved in the direction they did without claiming that any one strategy worked better than any other in improving the quality of American politics. It is not clear that any one did. If many Americans currently vote because of a prickly republican conscience, the promise of a highly participatory politics did not attract a significantly greater percentage of adult white males to the polls in the mid-1800s.[12] In the last analysis, the continuities in American political history stand out in bolder relief than the discontinuities. The nation's liberal tradition has not been univocal or all-inclusive, but it has been pervasive.

Notes

Chapter One

1. See Louis Hartz, *The Liberal Tradition in America* (New York: Harcourt, Brace, 1955).

2. See Bernard Bailyn, *The Ideological Origins of the American Revolution* (Cambridge, MA: Belknap, 1967). Other influential works of republican scholarship are: Gordon S. Wood, *The Creation of the American Republic, 1776–1787* (Chapel Hill: University of North Carolina Press, 1969); J. G. A. Pocock, *The Machiavellian Moment: Florentine Political Thought and the Atlantic Republican Tradition* (Princeton, NJ: Princeton University Press, 1975); Lance Banning, *The Jeffersonian Persuasion: Evolution of a Party Ideology* (Ithaca, NY: Cornell University Press, 1978).

3. See Joyce Appleby, "The Social Origins of American Revolutionary Philosophy," *Journal of American History* 64 (1977–78): 935–58. Also, Isaac Kramnick, "Republican Revisionism Revisited," *American Historical Review* 87 (1982): 629–64.

4. In his recent review article, Peter Onuf highlights this new aspect of the historiographical debate. See Peter S. Onuf, "Reflections on the Founding: Constitutional Historiography in Bicentennial Perspective," *William and Mary Quarterly*, 3d series, 46 (1989): 350–51, 353–54. Cf. Joyce Appleby, "Republicanism in Old and New Contexts," *William and Mary Quarterly*, 3d series, 43 (1986): 23–26; Lance Banning, "Jeffersonian Ideology Revisited: Liberal and Classical Ideas in the New American Republic," *William and Mary Quarterly*, 3d series, 43 (1986): 4, 12–14; Isaac Kramnick, "The 'Great National Discussion': The Discourse of Politics in 1787," *William and Mary Quarterly*, 3d series, 45 (1988): 3–32; Gordon S. Wood, "Ideology and the Origins of Liberal America," *William and Mary Quarterly*, 3d series, 44 (1987): 634.

5. To be useful, the definition should, of course, be generated from an analysis of particular historical cases.

6. See J. G. A. Pocock, "Machiavellian Moment Revisited: A Study in History and Ideology," *Journal of Modern History* 53 (1981): 70–71.

7. I will also offer fuller definitions of liberalism and pluralism in chapter 2. The "liberal universe" (or smorgasbord?) metaphor is meant not only to call attention to the common sources of American republicanism and pluralism but to emphasize that the process of transmission was an indirect one. Accordingly, I think it is a mistake to conceive of James Madison, for instance, as intending

to translate the political ideas of David Hume into a working public philosophy, however much he might have consciously drawn some of his arguments from him. Cf. Douglass Adair, "'That Politics May Be Reduced to a Science': David Hume, James Madison, and the Tenth Federalist," in *Fame and the Founding Fathers,* edited by Trevor Colbourn (New York: Norton, 1974), pp. 93–106. For a discussion of this methodological point, see Quentin Skinner, "The Limits of Historical Explanation," *Philosophy* 4 (1966): 199–215.

8. Actually, one or both of these caricatures are common to a number of different academic perspectives, not just the republican revisionists and neoliberals. Thus, while the Straussians, quite correctly, stress the distinction between classical and modern, liberal republicanism, they tend to lose sight of the fact that there can be, and have been in the course of American history, significant alternatives within liberalism. Esp., see Thomas L. Pangle, *The Spirit of Modern Republicanism: The Moral Vision of the American Founders and the Philosophy of Locke* (Chicago: University of Chicago Press, 1988), pp. 28–36. This narrow reading of liberalism is shared by philosophic communitarians and neo-Marxists. For example, see Charles Taylor, "Atomism," in *Philosophy and the Human Sciences* (Cambridge: Cambridge University Press, 1985), pp. 187–210; C. B. Macpherson, *The Political Theory of Possessive Individualism* (Oxford: Clarendon, 1962). What, then, is also at stake in this historiographical debate is whether liberalism can take (relatively) public-spirited forms. For the affirmative answer, see Nathan Tarcov, "A 'Non-Lockean' Locke and the Character of Liberalism," in *Liberalism Revisited,* edited by Douglass MacLean and Claudia Mills (Totowa, NJ: Rowman & Allanheld, 1983), pp. 130–40. Many of the historical fallacies behind the narrow view of liberalism are exposed in Stephen Holmes, "The Permanent Structure of Antiliberal Thought," in *Liberalism and the Moral Life,* edited by Nancy L. Rosenblum (Cambridge, MA: Harvard University Press, 1989), pp. 227–53.

9. Cf. Cicero, *The Laws* (Cambridge, MA: Harvard University Press, 1928), p. 337; Lester J. Cappon (ed.), *The Adams-Jefferson Letters* (Chapel Hill, NC: University of North Carolina Press, 1959), pp. 432–33, 437–38.

10. It is difficult to pin down the revisionist position on consensus. While the revisionists seem to agree that America exhibited a republican consensus in 1776, they diverge on how consensual it remained and for how long. Cf. Pocock, *The Machiavellian Moment,* chap. 15; Wood, *The Creation of the American Republic,* chap. 15. Here, I am following J. David Greenstone and Major Wilson in amending Hartz by positing different forms of liberalism within an overarching liberal consensus, though we each define those different forms of liberalism in distinctive ways. Cf. J. David Greenstone, "Political Culture and American Political Development: Liberty, Union, and the Liberal Bipolarity," *Studies in American Political Development* 1 (1986): 1–49; Major L. Wilson, *Space, Time, and Freedom: The Quest for Nationality and the Irrepressible Conflict* (Westport, CT: Greenwood, 1974), esp., chap. 1. Greenstone presents an expanded version of his counterthesis in his forthcoming book, *The Lincoln Persuasion; Polarity and Synthesis in American Politics* (Princeton, NJ: Princeton University Press, 1993), chap. 2.

11. Minimally, the claim is not that only a few are capable of articulating a coherent set of political ideas but that only a few do so. The seminal article remains: Philip E. Converse, "The Nature of Belief Systems in Mass Publics," in *Ideology and Discontent,* edited by David E. Apter (New York: Free Press, 1964), pp. 206–61. One indication of the changed nature of American political thought is the bifurcation of the role of statesman into politician and academic.

12. See John Ashworth, *'Agrarians' & 'Aristocrats': Party Political Ideology in the United States, 1837–1846* (Cambridge: Cambridge University Press, 1987), pp. 94–111; Marvin Meyers, *The Jacksonian Persuasion: Politics and Beliefs* (Stanford, CA: Stanford University Press, 1960), chap. 9. The relation of religious, particularly radical Protestant, thought to liberalism or liberal republicanism has generated a sizable body of literature. See Edmund S. Morgan, "The Puritan Ethic and the American Revolution," *William and Mary Quarterly,* 3d series, 24 (1967): 3–43; Sacvan Bercovitch, *The American Jeremiad* (Madison: University of Wisconsin Press, 1978); John P. Diggins, *The Lost Soul of American Politics* (Chicago: University of Chicago Press, 1984); James T. Kloppenberg, "The Virtues of Liberalism: Christianity, Republicanism, and Ethics in Early American Political Discourse," *Journal of American History* 74 (1987): 9–33.

13. The mystery is why the republican revisionists overlook both the fact that the major debates of the pre–Civil War period were over American nationality and how their own thesis can account for that fact.

14. I, of course, would not wish to deny that these debates possessed other dimensions, such as personal ambition, partisan advantage, and economic conflict. The participants, however, still had to interpret those dimensions, both to themselves and to other Americans. Thus, the question of public philosophy, and of differences between public philosophies, reasserts itself.

15. What may be controversial is treating Publius as one. The "split-personality" thesis, though, seems exaggerated. Cf. Douglass Adair, "The Authorship of the Disputed Federalist Papers," in *Fame and the Founding Fathers,* pp. 27–74; George W. Carey, "Publius—A Split Personality," *Review of Politics* 46 (1984): 5–22. The tensions between Alexander Hamilton's and Madison's *Federalist* papers are, in a significant sense, the tensions within their papers and their subsequent "divorce" does not obviate their 1787–88 commonalities. In any case, their pre- and post-*Federalist* differences are beyond my purview. The same is true of Calhoun's and Webster's manifold political gyrations before and after the nullification crisis. Fortunately or not, this interpretative problem does not arise for Federal Farmer because we are no longer sure who he was. (It had long been thought that he was Richard Henry Lee.) See Herbert J. Storing, *The Complete Anti-Federalist* (Chicago: University of Chicago Press, 1981), II:215–16; Gordon S. Wood, "The Authorship of the Letters from the Federal Farmer," *William and Mary Quarterly,* 3d series, 31 (1974): 299–308. But, cf. Robert H. Webking, "Melancton Smith and the Letters from the Federal Farmer," *William and Mary Quarterly,* 3d series, 44 (1987): 510–28. Finally, the choice of two Northern statesmen

debating the slavery issue, instead of a representative of each section, may require some justification. My justification is that the debates within each section were more historically important than the debates between the sections (which, in any case, had largely run their course by the late 1850s).

16. For all its virtues, Drew McCoy's study of the Jeffersonians is a good example of one which often lapses into this "save republicanism" approach. See Drew McCoy, *The Elusive Republic: Political Economy in Jeffersonian America* (Chapel Hill: University of North Carolina Press, 1980), pp. 9–10, 48–49, 236–37.

17. There is a rapidly growing body of literature about the essential (or contingent) contestability of political concepts. See Terence Ball, *Transforming Political Discourse: Political Theory and Critical Conceptual History* (Oxford: Basil Blackwell, 1988), chap. 1.

18. In a well-known series of articles, Quentin Skinner elaborated the distinction between these different types of historical explanation. These articles include: "Meaning and Understanding in the History of Ideas," *History and Theory* 8 (1969): 3–53; "Motives, Intentions and the Interpretation of Texts," *New Literary History* 3 (1972): 393–408; "'Social Meaning' and the Explanation of Social Action," in *Philosophy, Politics and Society,* edited by Peter Laslett (Oxford: Basil Blackwell, 1972), pp. 136–57; "Some Problems in the Analysis of Political Thought and Action," *Political Theory* 2 (1974): 277–303; "Hermeneutics and the Role of History," *New Literary History* 7 (1975): 209–32.

19. While Pocock has been criticized for making the unrealistic assumptions identified in the text, he does not, at least not according to a recent methodological essay. See J. G. A. Pocock, "The State of the Art," in *Virtue, Commerce, and History* (Cambridge: Cambridge University Press, 1985), pp. 1–34. Cf. Joyce Appleby, "Republicanism and Ideology," *American Quarterly* 37 (1985): 468–69; Kramnick, "The 'Great National Discussion,'" pp. 3–4. However, one of the reasons these critics might have thought he made those assumptions is precisely because his approach appears to lose much of its explanatory power without them.

20. In his major methodological statement, Wood explicitly recommends a mixed approach. See Gordon S. Wood, "Intellectual History and the Social Sciences," in *New Directions in American Intellectual History,* edited by John Higham and Paul K. Conkin (Baltimore: Johns Hopkins University Press, 1979), pp. 37–38. Yet, his *Creation of the American Republic* is bifurcated between analyses of the founding in terms of general intellectual trends and in terms of particularistic economic motives, and it is not at all clear how they are supposed to fit together. See Onuf, "Reflections on the Founding," pp. 349–50. Although in their more methodological discussions the revisionists uniformly claim to have transcended the debate over whether ideas are motives, their historical analyses do betray other tendencies. Cf. Bailyn, *The Ideological Origins of the American Revolution,* pp. 22, 95, 161, 231; Bernard Bailyn, "The Central Themes of the American Revolution: An Interpretation," in *Essays on the American Revolution,* edited by Stephen G. Kurtz and James H. Hutson (Chapel Hill: University of North Carolina Press, 1973), pp. 11, 23. Nonetheless, Diggins's iconoclastic critique that the expression of republican ideas is normally just rhetoric can, at best, only be

a partial truth. Is rhetoric ever just rhetoric? See Diggins, *The Lost Soul of American Politics*, esp., appendices.

21. Explaining precisely why Webster was a nationalist is, of course, where ulterior-motive accounts flourish. In Webster's case, the standard account is that his preference for a strong nation-state was primarily a matter of electoral politics since he represented an economically progressive state which benefited from interventionist federal policies. For example, see Richard N. Current, *Daniel Webster and the Rise of National Conservatism* (Boston: Little, Brown, 1955), pp. 192–93. This account, though, does not explain why his nationalism took a different form than that of other prominent Massachusetts politicians, such as John Quincy Adams.

22. It is somewhat arbitrary because American political thought was even then "adulterated" by pluralist (as well as nonliberal) ideas. However, the revisionists' argument that republican ideas had a firm grip on the minds of the American revolutionaries is persuasive. Their account of why is also persuasive. For obvious reasons, the American revolutionaries largely borrowed their arguments from the more republican, English opposition ("Country") polemicists, ignoring the more pluralistic, "Court" side of the political debates in the mother country. Consequently, what in England was a dialogue became a republican consensus in America. Esp., see Pocock, *The Machiavellian Moment*, pp. 506–9.

23. In parts two through five of *The Creation of the American Republic*, Wood breathtakingly describes an evolving tradition of political ideas in post-Revolutionary America. For him, though, it ultimately becomes not the story of an evolving intellectual tradition but of the death of American republicanism. His conclusion (part six) that the Anti-Federalists were residual republicans, while the Federalists were the forebearers of the liberal (pluralist) future, is jarring and drastically foreshortens the march of ideas in America. Post-*Creation*, both republican revisionists (including Wood himself) and their neoliberal critics have increasingly shown a greater sensitivity to the in-betweeness of historical cases. One consequence of this development has been the extension of the lifespan of American republicanism into the nineteenth century. Another consequence has been that the parties (Anti-Federalists, Jeffersonians, and Jacksonians) which initially looked more republican than their rivals (Federalists, National Republicans, and Whigs) now look less so. See Appleby, "Republicanism in Old and New Contexts," pp. 23–25; Ashworth, *'Agrarians' & 'Aristocrats,'* pp. 8–20, 52–61; Gordon S. Wood, "Interests and Disinterestedness in the Making of the Constitution," in *Beyond Confederation: Origins of the Constitution and American National Identity*, edited by Richard Beeman, Stephen Botein, and Edward C. Carter III (Chapel Hill: University of North Carolina Press, 1987), pp. 69–109. Yet, the two groups of scholars are still debating the relative influence (or mix) of classical and modern ideas in late-eighteenth- and early-nineteenth-century America rather than the state of one evolving liberal tradition. See Joyce Appleby, "Liberalism and Republicanism in the Historical Imagination," in *Liberalism and Republicanism in the Historical Imagination* (Cambridge, MA: Harvard University Press, 1992), pp. 1–33.

24. This relationship is usually expressed as the antithesis, or at least deep tension, between commerce and republicanism. As America became a more commercial nation, Americans were under increasing pressure to abandon republican ideas. Esp., see McCoy, *The Elusive Republic,* pp. 9, 48, 67, 236–37. This view is not so much false as too narrow.

25. Rowland Berthoff stresses the reactionary aspects of the nineteenth-century attempts to adjust the republican tradition to social change. See Rowland Berthoff, "Independence and Attachment, Virtue and Interest: From Republican Citizen to Free Enterpriser, 1787–1837," in *Uprooted Americans: Essays to Honor Oscar Handlin,* edited by Richard L. Bushman, Neil Harris, David Rothman, Barbara Miller Solomon, and Stephen Thernstrom (Boston: Little, Brown, 1979), pp. 99–124.

26. In this study, "nationalism" will refer to a political ideology (or, perhaps better, sensibility) which developed in America in the late eighteenth century. This ideology had *relatively* unambiguous structural implications and the interface between the two will become an important part of the argument of subsequent chapters. Historically, however, the ideology and the institutions were separable. See Hans Kohn, *The Idea of Nationalism: A Study in Its Origins and Background* (New York: Macmillan, 1946), chap. 1.

27. Wood notes how federalism was the uncontested future in 1776, in part because of its traditional association with republicanism. See Wood, *The Creation of the American Republic,* p. 95. He, however, ignores how American nationalism was a countervailing force—a force which had other roots in colonial experience (such as the colonists' growing sense of estrangement from the mother country); a force which must also be considered a significant cause (and effect) of the war of independence; and a force which had tremendous potential to upset the original federal-republican consensus. Cf. Onuf, "Reflections on the Founding," pp. 357–58.

28. The victory of American nationalism was true even in the South. As Arthur Bestor demonstrates, the sectional debates of the late 1850s were not between states' rights and nationalism but between two different versions of the latter. See Arthur Bestor, "State Sovereignty and Slavery: A Reinterpretation of Proslavery Constitutional Doctrine, 1846–1860," *Journal of the Illinois State Historical Society* 54 (1961): 117–80. The implication, though, is *not* that historically the states-rights position had merely been an intellectual scaffolding for whatever economic interests it seemed to serve. Cf. Arthur Schlesinger, Sr., "The State Rights Fetish," in *New Viewpoints in American History* (New York: Macmillan, 1922), pp. 220–44; Lewis O. Saum, "Schlesinger and 'The State Rights Fetish': A Note," *Civil War History* 24 (1978): 351–59.

29. While I naturally would not deny that prevailing ideas change quicker at some times than others, I suspect that not only Wood but also those who explicitly identify themselves as "conceptual change" theorists overestimate how rapidly, and how completely, it can occur. They are plainly wrong about the case of the American founding. Cf. Terence Ball and J. G. A. Pocock (eds.), *Conceptual Change and the Constitution,* (Lawrence: University Press of Kansas, 1988), esp., editors' introduction and articles by James Farr, Ball, and Russell Hanson. Ap-

pleby chides Pocock for straightforwardly applying Thomas Kuhn's understanding of scientific paradigms and revolutions to social history. See Appleby, "Republicanism and Ideology," p. 467.

30. See Paul C. Nagel, *This Sacred Trust: American Nationality, 1798–1898* (New York: Oxford University Press, 1971), p. 205.

31. See John P. Diggins, "Republicanism and Progressivism," *American Quarterly* 37 (1985): 573–76.

32. Cf. J. G. A. Pocock, "Virtue and Commerce in the Eighteenth Century," *Journal of Interdisciplinary History* 3 (1972): 134. This caution stems from the fact that almost everyone in this area of scholarship is uncomfortable with the relation between political thought and practice. Obviously, thought mattered; but how and how much? For example, see Pocock, "State of the Art," pp. 28–29. I will not propose any systematic answers to those questions in this study. However, it seems indisputable that the ratification, nullification, and slavery debates were three critical moments in American history when theory did strongly influence practice.

33. Dahl's dissatisfaction can clearly be traced through his corpus. Yet, he still is touting the superiority of his pluralist theory (polyarchy) over other contemporary models of democracy on grounds of feasibility. See Robert A. Dahl, *Democracy and Its Critics* (New Haven, CT: Yale University Press, 1989), pp. 298–301. Here, Dahl, without too much distortion, lumps republican revisionism together with Alasdair MacIntyre's neo-Aristotelianism. See Alasdair MacIntyre, *After Virtue* (Notre Dame: University of Notre Dame Press, 1984). He might equally have included contemporary models of participatory democracy. For example, see Benjamin Barber, *Strong Democracy: Participatory Politics for a New Age* (Berkeley: University of California Press, 1984). Perhaps the strongest statement of republican revivalism is William M. Sullivan, *Reconstructing Public Philosophy* (Berkeley: University of California Press, 1982).

34. For example, see Hannah Arendt, *On Revolution* (Harmondsworth, U.K.: Penguin, 1965), chap. 6. Arendt seems to be a largely unacknowleged source of republican scholarship. But, cf. Banning "Jeffersonian Ideology Revisited," pp. 17–19.

Chapter Two

1. If, for instance, Diggins's critique of the republican revisionists in *The Lost Soul of American Politics* profits from his amorphous usages of republicanism, his targets are partly to blame.

2. The following discussion draws heavily on part one of Wood's *Creation of the American Republic.* There, Wood implicitly develops an ideal type of republicanism from primary-source materials. My discussion is meant to supplement his, but also to more explicitly, and systematically, develop such an ideal type.

3. The political processes for defining and pursuing the public good will be considered below under the qualities of republican government.

4. See Lance Banning, "Some Second Thoughts on Virtue and the Course of Revolutionary Thinking," in *Conceptual Change and the Constitution,* pp. 194–212.

5. During the pre–Civil War period, the three major categories of exclusions from the citizen body were for race, sex, and (lack of) property. It was the first category which became the most contentious one during the period. American statesmen—North and South—did try to justify this exclusion on republican grounds, claiming that white self-government depended on black noncitizenship or slavery and that blacks were incapable of self-government anyway. See Edmund S. Morgan, "Slavery and Freedom: The American Paradox," *Journal of American History* 59 (1972): 5–29; Robert E. Shalhope, "Thomas Jefferson's Republicanism and Antebellum Southern Thought," *Journal of Southern History* 42 (1979): 3–26.

6. The famous exchange between Madison and Charles Pinckney in the Philadelphia Convention exposed the limits of this traditional social vision. See Max Farrand (ed.), *The Records of the Federal Convention of 1787* (New Haven, CT: Yale University Press, 1966), I:397–404, 410–12, 420–23.

7. As noted in chapter 1, American statesmen at first thought this corollary perfectly defined their geopolitical situation.

8. McCoy provides the best account of the attitudes of the founding generation toward commerce. He describes how most Americans did not desire a Spartalike, agricultural autarky but rather to enjoy what they saw as the positive consequences of modern commercial developments without being inflicted with the negative ones. They believed that this state of affairs was possible through low tariff rates (except on luxuries) and liberal land policies which would indefinitely maintain America as an "agricultural-market" republic. See McCoy, *The Elusive Republic,* pp. 83–84, 186, 236–37.

9. This confusion is certainly apparent in the secondary literature on the relation of republican to democratic government (and, more broadly, of republican to democratic thought). For the same reasons that we might wish to formulate an ideal type of republicanism, despite its essential contestabililty, we might wish to formulate an ideal type of republican (and democratic) government. This effort defines a markedly different approach from that taken, for example, in Russell L. Hanson, "'Commons' and 'Commonwealth' at the American Founding: Democratic Republicanism as the New American Hybrid," in *Conceptual Change and the Constitution,* pp. 165–93. But, cf. Robert W. Shoemaker, "'Democracy' and 'Republic' as Understood in Late Eighteenth-Century America," *American Speech* 41 (1966): 83–95.

10. It is beyond my present purposes to explore the many questions raised by these positions, such as: Who is an adult?; What is a "major" office?; How frequent are frequent elections?; What are the practical limits on the size of legislative bodies?

11. Although the plan of government Hamilton presented in Philadelphia fulfilled these conditions, it clearly was at the less democratic end of the continuum and rejected on those grounds. See Farrand, *The Records of the Federal Convention,* I:291–93.

12. In America, this experimentation was occurring on the state level throughout the 1770s and 1780s. Again, see Wood, *The Creation of the American Republic,* pts. II–IV.

13. The structural relationship between the American and English governments is a complicated one and it has been the subject of considerable debate in the secondary literature. The fact that the American governments were not based on hereditary social orders was crucial in convincing Americans that only their governments were republican ones. Still, strong similarities existed between the two models of government, similarities which permit us to call both models "mixed." Banning rightly insists on the persistence of mixed-government thought in America into the nineteenth century. See Banning, *The Jeffersonian Persuasion*, pp. 92–102.

14. Pocock's *Machiavellian Moment* emphasizes the currency of this restoration theme. Esp., see pp. 508, 519, 546–47. Yet, this "conservative" tendency must be counterbalanced with the experimental attitudes of the American founders. Cf. Bailyn, "The Central Themes of the American Revolution," p. 23.

15. It was in these terms that American statesmen legitimated their "Country" revolution against the English government. See Bailyn, *The Ideological Origins of the American Revolution*, chap. 4.

16. This core belief is treated here, instead of under the qualities of citizenship, because of its direct institutional implications.

17. Most of the state constitutions—Pennsylvania was the extreme case—violated the "rule of law" separation of powers by partially uniting legislative, executive, and judicial functions. This idea of a separation of powers developed before the "checks and balances" idea. During the American founding, these two ideas were not clearly distinguished from each other, nor from still older notions of mixed government. See David F. Epstein, *The Political Theory of 'The Federalist'* (Chicago: University of Chicago Press, 1984), pp. 130–38; Herbert J. Storing, *What the Anti-Federalists Were For* (Chicago: University of Chicago Press, 1981), pp. 59–63.

18. For example, cf. Pocock, *The Machiavellian Moment*, pp. 521, 527, 545–46; Kramnick, "Republican Revisionism Revisited," pp. 657–59.

19. See Niccolo Machiavelli, *The Prince* (Chicago: University of Chicago Press, 1985), p. 61.

20. Cf. Aristotle, *The Politics* (Chicago: University of Chicago Press, 1984), Bk. I, chaps. 8–11; John Locke, *Two Treatises of Government* (New York: New American Library, 1965), Bk. II, chap. 5.

21. If Machiavelli represents the more extreme, "utopian" critique, Rousseau represents the more evolutionary, "out-of-date" critique. For example, see Jean-Jacques Rousseau, *On the Social Contract* (New York: St. Martin's Press, 1978), pp. 102–3. My understanding of the relation between classical and modern liberal thought is indebted to Leo Strauss's works. Esp., see Leo Strauss, "What Is Political Philosophy?" in *What Is Political Philosophy? and Other Studies* (New York: Free Press, 1959), pp. 9–55.

22. In the following discussions of liberalism and pluralism, I will use the same four headings as I did in my discussion of republicanism but not the further notation. Differences between these categories of thought do not arise on all sixteen of the core beliefs of republicanism.

23. It is important to keep in mind that liberal statesmen can give these ends more or less public emphases.

24. See Thomas Hobbes, *Leviathan* (Harmondsworth, U.K.: Penguin, 1968), p. 188.

25. See Baron de Montesquieu, *The Spirit of Laws* (New York: Hafner, 1949), pp. lxxi, 23.

26. See Henry D. Aiken, ed., *Hume's Moral and Political Philosophy* (New York: Hafner, 1948), pp. 318, 384–85.

27. Eighteenth-century liberals considered the English government a liberal, but nonpopular, one. Compared to other European governments, however, it was relatively democratic. Montesquieu's discussion was, again, paradigmatic. See Montesquieu, *The Spirit of Laws,* Bk. XI, chap. 6.

28. See Arthur F. Bentley, *The Process of Government* (Chicago: University of Chicago Press, 1908); Charles Beard, *An Economic Interpretation of the Constitution of the United States* (New York: Macmillan, 1913). For a summary of this early-pluralist critique, see Paul F. Bourke, "The Pluralist Reading of James Madison's Tenth *Federalist,*" *Perspectives in American History* 9 (1975): 271–95. Not surprisingly, a division of labor developed at the time between historians (such as Beard) who criticized the nondemocratic past and political scientists (such as Bentley) who described the pluralistic present. But—and this is the point of Bourke's article—there also was a natural tendency among these scholars to discover precursors inside the pantheon of American statesmen. They often fastened on Madison's *Federalist* 10. According to Bourke, this quest led them to misinterpret Madison. Their misinterpretations will be noted in chapter 4.

29. The following discussion is indebted to Dahl's recent restatement of pluralist theory in *Democracy and Its Critics.* Esp., see pp. 215–23, 251–60, 278–91.

30. The pluralist analysis relied on new developments in the other social sciences which stressed the intrinsically social nature of human personality. For example, see George Herbert Mead, *Mind, Self and Society from the Standpoint of a Social Behaviorist* (Chicago: University of Chicago Press, 1934).

31. There is no ironclad rule on how much diversity is too much diversity even for pluralist societies. Such societies have, for instance, survived bipolar ethnic divisions for long periods of time through confederal constitutional arrangements.

32. This level-difference between pluralists and federal republicans will become extremely important in later chapters.

33. These criteria were discussed in the section on "Republicanism" under the qualities of republican government. The pluralists' stronger presumption in favor of private rights would only weaken this conclusion if, in contrast to republicans, they thought a less democratic government was more likely to protect those rights. This does not appear to be the case.

34. While any conception of politics contains a mixture of empirical and normative statements—with the one often masquerading as the other—pluralists clearly intend to be, and largely succeed in being, more empirical ("behavioristic") than their predecessors. Dahl's critique of Madisonian democracy is

paradigmatic here. See Robert A. Dahl, *A Preface to Democratic Theory* (Chicago: University of Chicago Press, 1956), chap. 1.

35. This loss of faith is really Wood's narrative in *The Creation of the American Republic.* As indicated in chapter 1, republican revisionists have subsequently drawn out the process, at least for some significant group of American statesmen.

36. Pocock's portrayal of their republicanism as a revolt against modernity seems particularly exaggerated. See Pocock, *The Machiavellian Moment,* pp. 509, 545–46.

37. Storing stresses how the Anti-Federalists had the weaker argument. See Storing, *What the Anti-Federalists Were For,* pp. 6, 71.

Chapter Three

1. "Observations Leading to a Fair Examination of the System of Government Proposed by the Late Convention, Letters from The Federal Farmer," in *The Complete Anti-Federalist,* II:229, letter I. [Hereafter cited in text and notes as: FF, with appropriate page and letter references.]

2. Previous interpretations of the ratification debate have only seen nationalism as an important factor on the Federalist side of the debate. Esp., see Stanley Elkins and Eric McKitrick, "The Founding Fathers: Young Men of the Revolution," in *The Reinterpretation of the American Revolution: 1763–1789,* edited by Jack P. Greene (New York: Harper & Row, 1968), pp. 381–82. However, the fact that the Anti-Federalists were not disunionists has been duly noted. See Jackson Turner Main, *The Antifederalists: Critics of the Constitution, 1781–1788* (New York: Norton, 1961), pp. 281–83. The claim that they were more politically cautious than the Federalists in the sense of being more traditional (even reactionary) statesmen has also been made. It, indeed, has formed the core of the three seminal (re)interpretations of the Anti-Federalists—as "men of little faith" (Cecelia Kenyon 1955); as "neoclassical republicans" (Wood 1969); and as "reluctant liberals" (Storing 1981). See Cecelia M. Kenyon, "Men of Little Faith: The Anti-Federalists on the Nature of Representative Government," in *The Reinterpretation of the American Revolution,* pp. 561–62, 565–66; Storing, *What the Anti-Federalists Were For,* pp. 6, 16, 46, 83 n.7; Wood, *The Creation of the American Republic,* pp. 514, 523, 560. (Wood has since significantly modified his view of the Anti-Federalists. See Wood, "Interests and Disinterestedness," pp. 93, 102, 109.) With the Federal Farmer as my test case, I would say that each, in her or his own way, slighted the Anti-Federalists' nationalism and the extent to which it "pulled" them beyond traditional political ideas.

3. Some of these amendments will be discussed in the course of this chapter. The Federal Farmer, though, presents neither a definitive list of amendments nor a consistent set of lists. He does not even insist on prior amendments, which quickly became the Anti-Federalists' rallying cry. See Robert A. Rutland, *The Ordeal of the Constitution: The Antifederalists and the Ratification Struggle of 1787–88* (Norman: University of Oklahoma Press, 1966), pp. 225, 236. [They were highly skeptical of the Federalist promise of amendments *after* ratification, although that promise did induce some of them to vote for the Constitution. The Federal

Farmer shares this skepticism (see pp. 250–51, letter IV).] Rutland, nonetheless, emphasizes the relatively mild nature of their opposition to the Constitution. His (partial) explanation is that the Anti-Federalists who are normally studied as representatives of the oppositional group were "elites" and, thus, not truly representative. See Rutland, *The Ordeal of the Constitution,* pp. 39–40. Main and Wood push much harder on this argument. See Main, *The Antifederalists,* pp. xi, 172–73, 177, 281; Wood, *The Creation of the American Republic,* p. 485; Wood, "Interests and Disinterestedness," p. 93. Yet, this argument could be made about the self-reputed spokesmen of any group or party. It fails to discriminate Anti-Federalists from Federalists. We also have no way of knowing whether many of the Anti-Federalist spokesmen, including the Federal Farmer, were elites other than the fact that their arguments against the Constitution were published in newspapers and were considered thoughtful ones both at the time and by later archivists and historians. Is not that exactly why we continue to study them as representatives of Anti-Federalist thought, regardless of their socioeconomic background?

4. Publius's only explicit reference to the Federal Farmer registers his approval of the constitutional provisions on presidential elections. See Clinton Rossiter (ed.), *The Federalist Papers* (New York: New American Library, 1961), p. 411, paper 68. [Hereafter cited in text and notes as: Publius, with appropriate page and paper references.] Murray Dry finds many other implicit references to each others' arguments, probably excessively so. See Murray P. Dry, "Anti-Federalism in *The Federalist:* A Founding Dialogue on the Constitution, Republican Government, and Federalism," in *Saving the Revolution: The Federalist Papers and the American Founding,* edited by Charles R. Kesler (New York: Free Press, 1987), pp. 40–60.

5. Starting with the publication of Storing's definitive collection of Anti-Federalist writings in 1981, a number of defenses of the coherence of Anti-Federalist thought have appeared. Esp., see Murray Dry, "The Case Against Ratification: Anti-Federalist Constitutional Thought," in *The Framing and Ratification of the Constitution,* edited by Leonard Levy and Dennis J. Mahoney (New York: Macmillan, 1987), pp. 271–91; Michael Lienesch, "In Defense of the Antifederalists," *History of Political Thought* 4 (1983): 65–87. Kenyon exemplified the prior view, which stressed the inconsistencies in Anti-Federalist thought. See Cecelia M. Kenyon (ed.), *The Antifederalists* (Indianapolis, IN: Bobbs-Merrill, 1966), pp. ci–cv.

6. Cf. Publius, pp. 516–17, paper 85.

7. Publius devotes paper 27 to discrediting the idea that the federal government will experience any special difficulties in enforcing its laws.

8. The Federal Farmer believes this despite Shays' Rebellion, which he treats seriously enough but still views as a small factional movement (see p. 253, letter V). Publius naturally believes that the rebellion had much broader implications (see p. 56, paper 6; pp. 166–67, paper 25; p. 178, paper 28; p. 448, paper 74). He also notes that most of the American states are much larger than small republics, at least as traditionally conceived (see p. 73, paper 9). While this observation might seem potentially devastating to his opponents' case, it remains

a dictum in his political argument. As we shall see in the next chapter, he concludes that the republican experience in the states is defective—and Shays' is one of his proofs—not because they are too large but because they are too small. For purposes of the ratification debate, both Publius and the Federal Farmer assume that the states are small republics.

9. Publius is more percipient about the possibilities for a federal administration of justice (see pp. 176–77, paper 27).

10. On the grounds that they will not demand the same amount of policing, the Federal Farmer distinguishes external (custom) from internal (excise) taxes and only criticizes federal powers over the latter.

11. To a lesser extent, the Federal Farmer also refers to the power of raising armies as an internal police power. Needless to say, his usages of "internal police" are not very consistent. The exact nature of his proposals on the sword and the purse will be discussed below.

12. Also, see FF, pp. 260–61, letter VI; p. 330, letter XVII; p. 341, letter XVIII. Although the Federal Farmer does not think that the condition of the country is rosy, he portrays it in a much less somber light than Publius does. He also claims that the dislocations of the war of independence, not the defects of the Articles of Confederation, are primarily responsible for whatever ills the nation is suffering. Cf. FF, pp. 224–26, letter I; p. 257, letter VI; p. 334, letter XVII; Publius, p. 59, paper 6; pp. 106–7, paper 15; p. 252, paper 40.

13. In this context, "mixed republic[s]" means not socially mixed but federal. As we will see, the Federal Farmer *mostly* does not conceive of the American republic as a socially mixed one, at least not in the traditional sense.

14. Also, see FF, p. 232, letter II; p. 236, letter III. The latter passage concerns the difficulties the convention faced in reconciling sectional differences. In it, the Federal Farmer describes the South as being ruled by a "dissipated aristocracy." (The South, then, is also his reference for the "less democratic" states. These passages are evidence for Wood that the Federal Farmer was not a Southerner, hence not Richard Henry Lee. See Wood, "Authorship of the Letters from the Federal Farmer," p. 302.) The Federal Farmer later says that the constitutional clause permitting Congress to prohibit the importation of slaves after 1808 is "not so favorable as could be wished for" (see p. 348, letter XVIII). While the statement is ambiguous, I presume that in light of his previous oblique references to slavery he supported an immediate prohibition. (This is not necessarily evidence that he was not Lee because Virginia had already prohibited the importation of slaves.) In sum, the Federal Farmer, as most other American statesmen at the time (including Publius), labored to *not* make slavery an issue during the ratification debate. See William W. Freehling, "The Founding Fathers and Slavery," *American Historical Review* 77 (1972): 81–93. For the relevant *Federalist* passages, see Publius, pp. 266–67, paper 42; 336–40, paper 54.

15. That is, this reason is sufficient whether or not the federal government will have to rely on force to execute its policies.

16. The Federal Farmer, though, does not really subscribe to the quoted theory of government even for small nations or states (see p. 248, letter IV; pp.

262–63, letter VI). Cf. Storing, *What the Anti-Federalists Were For,* pp. 56–58. But, then, neither did the Federalists defend such a theory.

17. I will not pursue the sufficient conditions, such as recall and mandatory rotation in office, which define the Federal Farmer's ideal government as even more strongly democratic (see pp. 289–92, letter XI; pp. 303–4, letter XIII). It is also beyond my present purposes to pursue his views on the Senate, presidency, and Supreme Court. In general, he finds those institutions well constructed in the Constitution (see pp. 287–88, letter XI; p. 310, letter XIV; p. 317, letter XV). When all is said and done, however, Kenyon's emphasis on the nondemocratic aspects of Anti-Federalist thought is as unbalanced as the overly democratic, Progressive interpretation she revised. Cf. Kenyon, "Men of Little Faith," pp. 528–29, 560–61; Beard, *An Economic Interpretation of the Constitution,* chap. 11. For a nicely balanced view, see Storing, *What the Anti-Federalists Were For,* pp. 39–40. The following discussion will show how closely intertwined the federal and democratic issues were for the Federal Farmer and, yet, how he saw the federal issue as the prior one. This priority has been reversed in revisionist interpretations of the Anti-Federalists (and Federalists). Cf. Wood, "Interests and Disinterestedness," pp. 72–73, 93, 101; Dry, "The Case Against Ratification," pp. 278–79.

18. We will discover that the Federal Farmer's usages of "orders" are also equivocal.

19. Here, Kenyon exaggerates the nondeliberative nature of the Anti-Federalist theory of representation. See Kenyon, *The Antifederalists,* pp. lix–lx, cviii. To stress the mirroring and sympathetic functions of representation is not to say that the representatives should not exercise some independent judgment when voting on laws.

20. The Federal Farmer also praises the House of Commons for its full and equal representation (see p. 272, letter VIII). Publius is much more critical of the English system of representation (see pp. 259–60, paper 41; p. 349, paper 56; p. 354, paper 57). But, cf. FF, p. 265, letter VII.

21. Also, see FF, pp. 242–43, letter III; p. 252, letter V; p. 282, letter X; p. 338, letter XVII.

22. Of course, the Federal Farmer's conspiratorial language is itself identifiably republican. See James H. Hutson, "Country, Court, and Constitution: Antifederalism and the Historians," *William and Mary Quarterly,* 3d series, 38 (1981): 362–63.

23. Also, see FF, p. 268, letter VII.

24. Notwithstanding the fact that the overwhelming majority is socially democratic, the Federal Farmer finds it inconceivable that aristocrats would ever not have their proportionate share of influence in government; the problem is always the reverse (see pp. 250–51, letter IV; p. 253, letter V; p. 267, letter VII; p. 272, letter VIII; p. 286, letter X). One source of confusion in the Federal Farmer's discussion of representation is his failure to sort out the relation he sees between social and natural aristocrats. He is well disposed toward the latter, which, again, is evidence of his republicanism (see p. 235, letter III; p. 267, letter VII).

25. The Federal Farmer goes on to apply the same criticism to Sparta. This passage belies any positive, classical influence on his political thought. Indeed, he appears to entirely dismiss such an influence when he observes that "we can put but little dependance [*sic*] on the partial and vague information transmitted to us respecting antient [*sic*] governments" and later that "we are now arrived to a new era in the affairs of men, when the true principles of government will be more fully unfolded than heretofore, and a new world, as it were, grow up in America" (see p. 260, letter VI; p. 286, letter X). For the two opposing views here, cf. Michael Lienesch, *New Order of the Ages: Time, the Constitution, and the Making of Modern American Political Thought* (Princeton, NJ: Princeton University Press, 1988), pp. 122–23; Gary L. Schmitt and Robert H. Webking, "Revolutionaries, Antifederalists, and Federalists: Comments on Gordon Wood's Understanding of the American Founding," *Political Science Reviewer* 9 (1979): 213–15.

26. The Federal Farmer even suggests that the federal representation will be more unequal than the Roman one. While there were "not more than 10 tribunes" at any one time in Rome (as compared to an initial 65 congressmen under the Constitution), they were annually, not biennially, elected and Rome was a smaller nation (see p. 273, letter VIII).

27. This fact also makes the English model of mixed government inappropriate for the American governments, requiring new justifications of bicameralism and a separation of powers (see p. 263, letter VI; p. 284, letter X; pp. 287–88, letter XI; p. 314, letter XIV). These passages stand as counterevidence to the thesis that such a redefinition only took hold on the Federalist side of the debate. See Lienesch, *New Order of the Ages*, pp. 131–32; Wood, *The Creation of the American Republic*, pp. 560–61.

28. In opposition to scholars who view the Anti-Federalists as agrarian democrats, it is significant that the Federal Farmer (despite his pseudonym) almost totally ignores any agrarian aspects to the ratification debate. (Conversely, why he calls himself the *Federal* Farmer and why his imaginary addressee is "The Republican" are immediately apparent to his readers.) He, in fact, seems very solicitous of the commercial classes. He considers their complaints about current economic conditions perfectly legitimate. He also commends them for normally being advocates of liberty. He even praises the "frugal merchants" of London, notwithstanding revisionist claims that Americans at the time tended to view London as a cesspool of finance capitalism (see p. 225, letter I; p. 252, letter V; p. 268, letter VII; p. 274, letter VIII). Cf. McCoy, *The Elusive Republic*, pp. 125–26. These passages suggest that the Federal Farmer had no difficulty envisioning members of the commercial classes as good citizens. [He, however, did have difficulty placing mechanics in that role, as traditionally would have been true of a republican statesman (see p. 268, letter VII).] Again, the retort is that Anti-Federalists like him were not true Anti-Federalists. See Main, *The Antifederalists*, pp. xi, 4–5, 166, 281.

29. The Bill of Rights, of course, was the Anti-Federalists' great victory, even if they were generally disappointed that it did not include stronger guarantees of state powers. See Dry, "The Case Against Ratification," pp. 287–89; Rutland,

The Ordeal of the Constitution, pp. 299–301; Storing, *What the Anti-Federalists Were For,* pp. 64–65.

30. The Federal Farmer does not have much faith that the federal representation will be increased after the first census, as provided in the Constitution (see p. 284, letter X). Publius belittles such concerns in paper 58. He also turns the tables on the Anti-Federalists, arguing that beyond a certain point larger representative bodies are actually less democratic ones. This is his early version of the iron law of oligarchy (see pp. 360–61, paper 58). The Federal Farmer, in contrast, defends relatively large representative bodies (see p. 284, letter X).

31. The departure on the sword and the purse depends on a somewhat unusual analogy between the American and English political systems in which a nondemocratic federal government represents the crown and the democratic state governments represent the House of Commons. It is noteworthy that the Federal Farmer ultimately rejects this analogy, preferring a federal government which not only combines some sword and purse powers but is as democratic as possible. Cf. Dry, "The Case Against Ratification," pp. 278–79.

32. Also, see FF, p. 260, letter VI; p. 265, letter VII; p. 273, letter VIII; p. 286, letter X. According to the Federal Farmer, it is primarily the scale of the American experiment that requires his countrymen to transcend their English precedents (see p. 260, letter VI; p. 274, letter VIII; p. 282, letter X; p. 335, letter XVII).

33. The tendency is, again, to only credit the Federalists with these intellectual moves. For example, see Wood, *The Creation of the American Republic,* pp. 499–500, 526–27.

34. Unfortunately, the Federal Farmer never distinguishes those powers which he would confer on the federal government only after its representation has been increased and those which he would confer on it "as is."

35. By "structurally," I mean the structure of the federal system, not the internal structure of the federal government. With respect to the latter, the Federal Farmer considers the Constitution superior because of the lack of a separation of powers in the Articles (see pp. 283–84, letter X). Even in this passage, though, he couples structure with powers, which always seems to be the prior issue for him. Publius would reverse that emphasis (see p. 156, paper 23; p. 196, paper 31). Cf. Storing, *What the Anti-Federalists Were For,* pp. 53–54.

36. For a judicious assessment of the degree to which the party labels were, and were not, misnomers, see Storing, *What the Anti-Federalists Were For,* pp. 9–10.

37. The same ambiguity appears in both the Federal Farmer's letters and Publius's papers between redefining federalism to fit their preferred plans and admitting that, as traditionally defined, their preferred plans are only partly federal ones. Cf. Publius, pp. 75–76, paper 9; pp. 243–46, paper 39. Insofar as there was an extension of the language from traditional (con)federalism to a new federalism of divided and independent powers, as Martin Diamond argued, Anti-Federalists as well as Federalists participated in that extension. See Martin Diamond, "What the Framers Meant by Federalism," in *A Nation of States,* edited by Robert A. Goldwin (Chicago: Rand McNally, 1961), pp. 24–41. Storing shows how the Anti-Federalists did eventually embrace this redefinition of federalism.

See Storing, *What the Anti-Federalists Were For,* pp. 33, 37. In the Federal Farmer's case, however, his initial position was not confederalism but consolidation; nor, as we shall see, was his final position the dual federalism Storing ascribes to the Anti-Federalists.

38. Equal state suffrage was partially embodied in the new federal government through the Senate. Ironically, the Federal Farmer at first objected to this feature of the Constitution on the grounds that it was undemocratic (because not proportionate to population), just as Madison, among others, had done at the convention. But the Federal Farmer later commended equal state representation in the Senate as proper to a plan of partial consolidation, while Madison as Publius could still only give it grudging support. Cf. Farrand, *The Records of the Federal Convention,* I:151–52, 485–87; FF, pp. 236–37, letter III; p. 287, letter XI; Publius, pp. 377–78, paper 62.

39. A somewhat less constant refrain is that the Constitution's grants of powers are poorly defined (see FF, p. 234, letter III; pp. 245–47, letter IV; p. 325, letter XVI; p. 340, letter XVII).

40. For conflicting statements on the desirability of a federal bankruptcy power, see FF, p. 229, letter I; p. 243, letter III; pp. 343–44, letter XVIII.

41. In particular, the Federal Farmer does not criticize the interstate commerce clause or the constitutional restrictions on state issues of paper money and impairment of contracts, provisions which Beard identified as central to the Anti-Federalists' democratic-agrarian critique of the Constitution. See Beard, *An Economic Interpretation of the Constitution,* pp. 292–93. Main and Wood would, once more, explain this "failure" through the distinction between elite and mass Anti-Federalists. See Main, *The Antifederalists,* pp. 166–67; Wood, "Interests and Disinterestedness," pp. 107–8. Cf. Kenyon, *The Antifederalists,* pp. lxxxvi–xci.

42. Publius also argues that the difference in modes is crucial. Yet, he does so sometimes to emphasize the differences between the Articles and the Constitution and sometimes to minimize them. (After all, the "powers" of the federal government are similar!) For example, cf. pp. 153–55, paper 23; p. 293, paper 45.

43. The Federal Farmer's only reference to other federations is in the context of requisition systems, where he mentions modern Germany and the Amphictyonic council of ancient Greece (see pp. 331, 333, letter XVII). Publius's references are much more extensive and negative. Esp., see pp. 122–38, papers 18–20. With regard to the Netherlands, he claims that only the extraordinary powers of the ruler of the dominant province (Holland) hold the federation together (see p. 137, paper 20).

44. For a similar expression of the idea of an internal balance of power, see FF, pp. 308–9, letter XIV. The Federal Farmer, therefore, provides counterevidence to another one of Kenyon's theses: that the Anti-Federalists favored a strict separation of powers. See Kenyon, *The Antifederalists,* pp. lxxvi–lxxx. Cf. Storing, *What the Anti-Federalists Were For,* pp. 55–63.

45. Although Publius agrees that the Articles exposed the federal government to the charity of the states governments, he vehemently denies that the Constitution leaves the state governments defenseless (see p. 111, paper 15; p.

119, paper 17; pp. 180–81, paper 28; pp. 289–90, paper 45; pp. 296–98, paper 46). But I fear that he doth protest too much.

46. Such a coercive requisition system—along with a federal commerce power—is often presented as the Anti-Federalists' preferred way of reforming the Articles. For example, see Main, *The Antifederalists,* p. 184.

47. Publius denounces supermajority provisions in the usual terms, as undemocratic and tending toward anarchy (see pp. 147–48, paper 22).

48. Publius is incredulous that concurrent taxing powers should be considered so problematic (see p. 200, paper 32; pp. 206–7, paper 34; p. 221, paper 36).

49. "Logical inference" refers to the Federalist argument that the federal government's means should be proportioned to its ends; for example, that it should have sufficient taxing powers to fund an adequate national defense. See Publius, pp. 153–54, paper 23; pp. 193–95, paper 31.

50. Cf. Publius, p. 190, paper 30; pp. 195–96, paper 32; pp. 207–8, paper 34.

Chapter Four

1. Although Publius, in his one explicit reference to the Federal Farmer, calls him "the most plausible" of the opponents of the Constitution, he is the most plausible of a group Publius variously describes as: "visionary"; attempting "to reconcile contradictions"; expecting "halcyon scenes of a poetic and fabulous age"; "specious"; beset by "chimerical fears"; and relying "merely on verbal and nominal distinctions." (And this is only a small sample!) See Publius, p. 56, paper 6; p. 157, paper 23; p. 192, paper 30; p. 300, paper 46; p. 411, paper 68; p. 515, paper 84.

2. In view of my claim that the federal issue was the prior one during the ratification debate, I will deemphasize the fourth head (papers 37–85). The basic point of those papers—that the proposed government is a republican one—will be important to the argument of this chapter but not the detailed defense of its internal structure. (Publius does not explicitly discuss the fifth and sixth heads outlined at the end of the first paper, except cursorily in the last paper.)

3. As we saw in chapter 1, these were the presumptions of republican revisionism, at least in its early versions and once we redefine the terms of debate from classical republicanism and modern liberalism to republicanism and pluralism. For example, see Wood, *The Creation of the American Republic,* pp. 499–500, 523–24, 560–64, 606–15.

4. The thesis that Publius's professed commitment to federalism was only (mostly?) intended to mollify the opponents of the Constitution hardly needs argument. Diamond wrote the seminal article in this reinterpretation of Publius's intentions. See Martin Diamond, "*The Federalist's* View of Federalism," in *Essays in Federalism* (Claremont, CA: Institute for Studies in Federalism, 1961), pp. 21–64. But, cf. Michael P. Zuckert, "A System Without Precedent: Federalism in the American Constitution," in *The Framing and Ratification of the Constitution,* pp. 132–50.

5. For the distinction between the union as a means and as an end, esp., see Paul C. Nagel, *One Nation Indivisible: The Union in American Thought, 1776–1861*

(New York: Oxford University Press, 1964). Nagel, however, employs this distinction to artificially divide the first and third generations ("means") under the Constitution from the second generation ("end") when the disagreements, both across and within generations, were more over the nature of the union as a means than as an end. See Nagel, *One Nation Indivisible,* pp. 14–22.

6. It has become fairly common to interpret Publius as a liberal hybrid; but as a classical republican-modern liberal hybrid, not as a republican-pluralist one. For example, see Wood, "Interests and Disinterestedness," p. 92.

7. Without trying to settle the imbroglio over how property-conscious Publius and other Federalists were, and whether they were more so than the Anti-Federalists, I would simply point out that the Federal Farmer, in the opening paragraph of his letters, announces his commitment to the protection of property (see p. 224, letter I). For secondary discussion, see James H. Hutson, "The Constitution: An Economic Document," in *The Framing and Ratification of the Constitution,* pp. 259–71.

8. Publius also uses these terms interchangeably throughout his other papers.

9. Significantly, the Federal Farmer fears the same possibility (see p. 224, letter I).

10. Not surprisingly, the Federal Farmer is much more troubled by the threat of minority, than majority, tyranny (see p. 269, letter VII).

11. These are probabilistic principles, for Publius has already discounted the possibility of enlightened statesmen always being at the helm (see p. 80, paper 10). Cf. Robert J. Morgan, "Madison's Theory of Representation in the Tenth Federalist," *Journal of Politics* 36 (1974): 852–85.

12. It is *almost* by definition because it is possible that a democratic assembly could be led by enlightened statesmen (such as Pericles?). Publius, though, dismisses this possibility in two oft-quoted passages on ancient Athens, his primary reference for a pure democracy (see p. 342, paper 55; p. 384, paper 63). Earlier, he had obliquely attacked Pericles's statesmanship (see pp. 54–55, paper 6).

13. In paper fifty-one, Publius adds the important proviso that majorities will still form in extended republics but seldom "on any other principles than those of justice and the general good"—that is, not as factions (see p. 325, paper 51).

14. Federalism does play some role in both papers ten and fifty-one. In paper ten, Publius praises federalism because it permits the superintendence of national interests by the federal government and local interests by the state governments (see p. 83, paper 10). This state of affairs, however, need be nothing more than administrative decentralization. In paper fifty-one, Publius claims that federalism allows for the possibility of extended republics (see p. 325, paper 51). Yet, in paper ten he claims that the representative principle does (see p. 83, paper 10). (As we shall see below, this tension reappears in paper fourteen.) The main point, though, is that a strong commitment to federalism would undercut the rhetorical force of the argument of these papers, which is to make the case for large-scale over small-scale governments of equal powers. Again, cf. Diamond, "*The Federalist's* View of Federalism," pp. 59–62; Zuckert, "System Without Precedent," p. 146.

15. For example, see James Conniff, "On the Obsolescence of the General Will: Rousseau, Madison, and the Evolution of Republican Political Thought," *Western Political Quarterly* 28 (1975): 52–55.

16. For example, see Martin Diamond, "Democracy and *The Federalist:* A Reconsideration of the Framers' Intents," in *The Reinterpretation of the American Revolution,* p. 518.

17. In the course of a very close analysis of *The Federalist Papers,* Epstein argues that Publius put at least equal weight on private and public liberty. See Epstein, *The Political Theory of 'The Federalist,'* pp. 5–7, 124–25, 147–49. For Epstein, this is evidence of the hybrid nature of Publius's political thought.

18. If my view of what Publius meant by the public good (and variants) is correct, then it is undefinable except in concrete cases, as in his statement that relying on tariffs for public revenues promotes "our political welfare" (see p. 93, paper 12). Of course, this lack of definition also makes it impossible to refute the counterargument that his meaning was the more pluralistic one of merely protecting rights and balancing interests. Cf. Edward J. Erler, "The Problem of the Public Good in *The Federalist,*" *Polity* 13 (1981): 649–67; Schmitt and Webking, "Revolutionaries, Antifederalists, and Federalists," pp. 197–212.

19. Dahl shows just how difficult it is to fit Publius's definition of factions into a pluralistic (and behavioristic) paradigm of politics. See Dahl, *A Preface to Democratic Theory,* pp. 25–26. Also, Bourke, "Pluralist Reading of James Madison's Tenth *Federalist,*" pp. 294–95. The Federal Farmer seems to share Publius's anti-party bias (see p. 253, letter V).

20. In thinking about Publius as a republican-pluralist hybrid, Kramnick's categorization of the dispersement principle as pluralistic and the refinement principle as republican is appealing. See Kramnick, "The 'Great National Discussion,'" pp. 6, 12–13. Publius, however, did not himself separate the two principles into distinct social and political visions, a point Epstein makes in criticizing the common tendency to emphasize one principle at the expense of the other. See Epstein, *The Political Theory of 'The Federalist,'* pp. 108–9. Both Kramnick and Wood (post-*Creation*) argue that the Anti-Federalists were, in some respects, more pluralistic than the Federalists; both cite the Anti-Federalist theory of a full and equal representation as evidence. See Kramnick, "The 'Great National Discussion,'" pp. 14–15; Wood, "Interests and Disinterestedness," pp. 101–2. This argument is a major contribution to the literature. Yet, Wood, at least, now clearly exaggerates the Anti-Federalists' pluralism, as in overlooking how they still clung to traditional republican assumptions on the state level. Overall, the Federalists were the more pluralistic group, even if they, too, remained mostly republican.

21. Adair wields both criticisms against the Progressive historians. See Douglass Adair, "The Tenth Federalist Revisited," in *Fame and the Founding Fathers,* pp. 75–76, 90–91.

22. As we saw in chapter 3, the Federal Farmer accepts this underlying harmony of interests between American agriculture and commerce. Publius, though, does devote much more effort to presenting the positive case for commerce. Without offering more proof here, I would argue that his case was pri-

marily a republican one. Cf. Martin Diamond, "The Federalist," in *History of Political Philosophy,* edited by Leo Strauss and Joseph Cropsey (Chicago: University of Chicago Press, 1987), pp. 668–70; Charles R. Kesler, "*Federalist* 10 and American Republicanism," in *Saving the Revolution,* p. 30.

23. Also, see Publius, p. 143, paper 21; p. 222, paper 36; p. 292, paper 45.

24. While the Federal Farmer and Publius agree in their empirical theories of representation, they sharply disagree in their normative theories. In particular, Publius is much more favorable to the probability that lawyers will dominate Congress. See Jean Yarbrough, "Representation and Republicanism: Two Views," *Publius* 9 (1979): 77–98.

25. In chapter 3, we noted that the Federal Farmer did not disregard how the American people were ideologically homogeneous and how that might allow for a stronger federal government. But, more, he formulated the apparent *lack* of *economic* homogeneity on the national level into a powerful argument for state powers.

26. Also, see Publius, p. 43, paper 3; p. 174, paper 27; p. 354, paper 57.

27. See Daniel Walker Howe, "The Language of Faculty Psychology in *The Federalist Papers,*" in *Conceptual Change and the Constitution,* pp. 116–22. Howe contends that the Anti-Federalists, in contrast, were more trusting of mass than elite virtue, as is apparent in the case of the Federal Farmer. Cf. Howe, "The Language of Faculty Psychology," pp. 119, 127–28; FF, p. 234, letter III; p. 266, letter VII; pp. 302–4, letter XIII.

28. Not surprisingly, Publius does not always express confidence in the Anti-Federalists' patriotism (see p. 34, paper 1; p. 74, paper 9; pp. 151–52, paper 22).

29. Publius, again, personalizes this fault in his opponents (see p. 35, paper 1; p. 356, paper 58; p. 387, paper 63).

30. The Federal Farmer, conversely, hesitated at a transfer of powers to the federal level precisely because the federal government would be less participatory than the state governments. The contrast with Publius here shows just how closely connected the federal and democratic issues were during the ratification debate.

31. Although at the start of the fourth head Publius demonstrates how the federal government will be at least as participatory as one of the state governments on each major office, he also betrays how, as a whole, it will be less participatory than every one of the state governments (see pp. 241–42, paper 39).

32. When first broaching the idea of auxiliary precautions, Publius also emphasized that they were auxiliary to popular elections (see p. 322, paper 51). He, thus, effectively turns the tables on the Anti-Federalists, claiming that it is really they, not the Federalists, who do not trust the people. The issue, however, was not whether most Americans could be trusted to choose good rulers. It was in what sort of situations were they more likely to choose good rulers. (This issue was, in turn, complicated by shifting definitions of "good rulers.") In substantiating the Anti-Federalists' nondemocratic tendencies, Kenyon is too eager to accept the Federalists' characterization of the "trust" issue. See Kenyon, "Men of Little Faith," pp. 556–58. The Anti-Federalists were the more democratic group, as evidenced by the opposing views of the Federal Farmer and Publius

on majority tyranny, social equality, political representation, civic virtue, and popular participation. This conclusion remains true even if we agree, as I think we should, that the Anti-Federalists were not "latter-day" democrats and that their differences with the Federalists were differences in degree, not kind. For an excellent discussion of these differences, see Storing, *What the Anti-Federalists Were For,* pp. 39–45. Diamond, of course, is the great defender of Publius's "enlightened" commitment to democracy. See Diamond, "Democracy and *The Federalist,*" pp. 510–16.

33. This case may not seem as revealing for purposes of giving an account of the ratification debate since it was Jefferson, not an Anti-Federalist, who made the proposal. Jefferson's proposal is, nonetheless, imbued with an Anti-Federalist spirit, as indicated by the Federal Farmer's similarly motivated proposals for recall and mandatory rotation in office. Jefferson, moreover, only supported the Constitution with strong reservations, just as some of the Anti-Federalists eventually did. See Thomas Jefferson, "Letter to James Madison" (Dec. 20, 1787), in *The Portable Thomas Jefferson,* edited by Merrill D. Peterson (Harmondsworth, U.K.: Penguin, 1977), pp. 428–33.

34. Publius also criticizes the proposal because it is unlikely to produce good solutions to "separation of powers" disputes (see pp. 315–17, paper 49).

35. The possibility of divided loyalties exposes the ambiguous meaning of patriotism. I will continue using "patriotism" to refer to the national variety, reserving "localist sentiments" for the subnational (most importantly, state) varieties. Most scholars would argue that patriotism or nationalism was the weaker sentiment at the time, except among the Federalist elite. For example, see Kohn, *The Idea of Nationalism,* pp. 287–88. This assessment may well be true. Still, it should not conceal the inroads nationalism had made among both the Anti-Federalist elite and the general populace. Rutland, for instance, notes how localist sentiments constituted a double-edged sword for the Anti-Federalists but not how their own nationalism partially explains their hesitancy to wield it. See Rutland, *The Ordeal of the Constitution,* p. 313.

36. Also, see Publius, p. 334, paper 53.

37. Also, see Publius, pp. 290–95, papers 45–46.

38. Publius is scathing in his criticisms of how the state governments have heretofore been administered (see p. 65, paper 7; pp. 77–78, paper 10; p. 227, paper 37; pp. 281–83, paper 44; pp. 521–22, paper 85). The Federal Farmer is also critical, but much less so (see pp. 225–27, letter I; p. 252, letter V; p. 276, letter IX; pp. 332–34, letter XVII). Actually, Publius foreshadows his later argument in paper seventeen. He predicts that the powers of the state governments will be too insignificant to the leaders of the federal government for them to want to usurp those powers and that the state governments will command their advantage in physical proximity "unless the force of that principle should be destroyed by a much better administration" of the federal government (see pp. 118–19, paper 17).

39. In offering a formal definition of republican government, Publius claims that any such definition must be fairly flexible as to manner of election and tenure. Otherwise, republican government not only would become a null

set but it would, by definition, be precluded from being a well-administered government (see p. 241, paper 39). Somewhat ironically, both he and the Federal Farmer quote Alexander Pope on the importance of administration to good government. Cf. Publius, p. 414, paper 69; FF, p. 224, letter I. The Federal Farmer, though, would distinguish himself from Publius here by (1) predicting that the state governments will be the better-administered ones; (2) pairing good administration with closeness to the people; (3) stressing closeness to the people as a partially separate criterion; and (4) insisting that the federal government can never develop this closeness even in a psychological sense.

40. Also, see Publius, pp. 71–72, paper 9; pp. 81, 83–84, paper 10. When he takes up the subject for the last time in paper sixty-three, Publius amends previous discussions by conceding that the Greek city-states did possess representative institutions. Their real deficiency was their compactness (see pp. 385–87, paper 63). It seems indisputable that Publius, notwithstanding his pseudonym, did not put much stock in classical political arrangements, except for their admonitory value (also, see p. 57, paper 6; pp. 68–69, paper 8; pp. 122–28, paper 18; pp. 232–33, paper 38; p. 342, paper 55; p. 425, paper 70). The fundamentally modern (liberal) nature of Publius's political thought is another one of Diamond's major themes. See Diamond, "Democracy and *The Federalist*," pp. 516–20. But, cf. Kesler, "*Federalist* 10 and American Republicanism," pp. 18–19, 22–23.

41. Publius's formal definition of republican government excludes these European nations (Germany, Poland, France, Spain, and England) because of their hereditary institutions.

42. Publius does not wonder at his predecessors' errors on the federal level due to the lack of good precedents, as he is so anxious to demonstrate in the next head.

43. For example, see Publius, p. 57, paper 6; p. 290, paper 45; p. 327, paper 52; p. 385, paper 63; pp. 526–27, paper 85. Cf. Douglass Adair, "'Experience Must Be Our Guide': History, Democratic Theory, and the United States Constitution," in *Fame and the Founding Fathers,* pp. 108–23.

44. Even though both appeal to the American sense of mission, the Federal Farmer is relatively more reverential toward "the opinions of former times and other nations." See Lienesch, *New Order of the Ages,* pp. 121–30.

45. For the rapid disappearance of Anti-Federalism, see Lance Banning, "Republican Ideology and the Triumph of the Constitution, 1789–1793," *William and Mary Quarterly,* 3d series, 31 (1974): 167–88. But, cf. Lienesch, *New Order of the Ages,* pp. 159–60.

Chapter Five

1. Only the two senators from South Carolina (Calhoun and Stephen Miller) defended their state's actions in nullifying the federal tariffs of 1828 and 1832, although five other Southern senators (George Bibb, D-Kentucky; Bedford Brown, D-North Carolina; Gabriel Moore, D-Alabama; George Poindexter, D-Mississippi; and John Tyler, D-Virginia) were very sympathetic to South Caro-

lina's plight and articulated constitutional doctrines which were nearly as extreme as the one upon which she had acted. The nullification debate was precipitated by the first reading of the Force Bill in the Senate on 28 January 1833 and it essentially ended with the passage of the bill on 20 February. The debate is reported in *Register of Debates in Congress* (Washington: Gales & Seaton, 1833) IX:244–688. [Hereafter cited in text and notes as *Debates,* with appropriate references to volume, page, and speaker.] The Force Bill authorized President Jackson to use federal troops, if necessary, to collect the tariff duties in South Carolina. Neither the Force Bill nor South Carolina's nullification ordinances actually went into effect as the nullification crisis was defused by the passage of Henry Clay's (NR-Kentucky) compromise tariff bill on 1 March. (In referring to senators, "D" and "NR" stand for, respectively, Democrat and National Republican, the two competing parties in 1833.)

2. See John C. Calhoun, *Works* (New York: Appleton, 1851–57) II:252–56. (Hereafter cited in text and notes as Calhoun, with appropriate volume and page references.) Calhoun delivered two nullification speeches in the Senate, "On the Revenue Collection [Force] Bill" (15–16 February; II:197–262) and "In Reply to Mr. Webster" (26 February; II:262–309). The first was a wide-ranging defense of his state's actions; the second was a more directed response to Webster's Senate speech of 16 February attacking nullification doctrine. See Daniel Webster, *Writings and Speeches* (Boston: Little, Brown, 1903), VI:181–238 ("The Constitution not a Compact between Sovereign States"), hereafter cited in text and notes as Webster, with appropriate volume and page references. For the saliency of the restoration theme in Jacksonian politics, see Meyers, *The Jacksonian Persuasion,* chap. 2.

3. The difference is that Calhoun's doctrine would arrest the assertion of a disputed federal power in the nullifying state until the Constitution was amended to explicitly grant that power, while the Federal Farmer's checks would allow the assertion of that power until it was vetoed by state governments which collectively represented a majority of the American people. Calhoun's understanding of American history is elaborated in *A Discourse on the Constitution and Government of the United States* (I:111–406), which he wrote during the 1840s and which was published, shortly after his death, in 1851. That work is less charitable to the founders. For example, it traces consolidationist declension back to *The Federalist Papers;* his nullification speeches date that trend from *McCullouch* v. *Maryland* (1819), sliding over Madison's convention actions and *Federalist* papers for his Virginia Resolutions of 1798–99 and even absolving Secretary of the Treasury Hamilton of consolidationist tendencies (cf. I:158–61, 341; II:298–99, 303).

4. Cf. Hartz, *The Liberal Tradition in America,* p. 9; Major L. Wilson, "Liberty and Union: An Analysis of Three Concepts Involved in the Nullification Controversy," *Journal of Southern History* 33 (1967): 331.

5. Calhoun emphasizes that his doctrine would leave the federal government stronger than it had been under the Articles of Confederation because it could still act without the prior consent of the state governments. States then would have to actively oppose federal laws, instead of merely fail to act on them (see II:304–5).

6. For strong statements of the two opposing views here, cf. Peter F. Drucker, "A Key to Modern Politics: Calhoun's Pluralism," *Review of Politics* 10 (1948): 412–26; J. William Harris, "Last of the Classical Republicans: An Interpretation of John C. Calhoun," *Civil War History* 30 (1984): 255–67. Harris, in line with the general tendencies of republican revisionism, not only exaggerates the classical elements in Calhoun's political thought but sees Calhoun's republicanism as consisting primarily in his theory of mixed government instead of his theory of federalism. For more balanced views of Calhoun's place within the American liberal tradition, see Lacy K. Ford, "Republican Ideology in a Slave Society: The Political Economy of John C. Calhoun," *Journal of Southern History* 54 (1988): 405–24; Ralph Lerner, "Calhoun's New Science of Politics," *American Political Science Review* 57 (1963): 918–32; Peter J. Steinberger, "Calhoun's Concept of the Public Interest: A Clarification," *Polity* 13 (1981): 410–24. Although a consensus exists among his biographers that Calhoun was a nationalist, they are not quite sure what to make of that fact. The problem is often solved temporally, as the early nationalist is sharply distinguished from the late sectionalist. The titles of Charles Wiltse's multivolume biography reflect this solution. See Charles M. Wiltse, *John C. Calhoun: Nationalist, 1782–1828; John C. Calhoun: Nullifier, 1829–1839; John C. Calhoun: Sectionalist, 1840–1850* (Indianapolis: Bobbs-Merrill, 1944–51). I would suggest another solution to this interpretive problem: Calhoun was always committed to the union, but at different stages of his career he appeared more or less nationalistic depending on how his other ideological and political commitments impinged on his commitment to the union. Cf. Pauline Maier, "The Road Not Taken: Nullification, John C. Calhoun, and the Revolutionary Tradition in South Carolina," *South Carolina Magazine* 82 (1981): 12.

7. The federal thrust of Calhoun's doctrine is evident from his rejection of the Senate, internal checks and balances, and the representative principle as sufficient safeguards against majority tyranny within the federal government (see II:256–57, 306). For now, my point is only that his desire to limit *national* majorities is not sufficient evidence that his political position was undemocratic (or unrepublican). For opposing views of Calhoun's commitment to democracy, cf. Richard Hofstadter, *The American Political Tradition and the Men Who Made It* (New York: Knopf, 1948), chap. 4; Charles M. Wiltse, "Calhoun's Democracy," *Journal of Politics* (1941): 210–23.

8. Also, see Calhoun II:225, 308–9. William Freehling's narrative of the crisis places the most weight on the slavery issue. See William W. Freehling, *Prelude to Civil War: The Nullification Controversy in South Carolina* (New York: Harper & Row, 1968), esp., pp. x, 49–52, 85–86, 257–59.

9. See *Debates* IX:490–91 (Moore). The other explicit references to slavery are carefully couched as hypotheticals; for instance, if South Carolina can nullify federal tariff laws, can not Pennsylvania nullify federal fugitive-slave laws? See *Debates* IX:261–62 (William Wilkins, D-Pennsylvania); 356 (John Holmes, D-Maine).

10. See Richard E. Ellis, *The Union at Risk: Jacksonian Democracy, States' Rights, and the Nullification Crisis* (New York: Oxford University Press, 1987), pp. 191–94. Calhoun's 1848 Senate speech in opposition to a bill organizing Oregon as

a free territory is probably his most strident proslavery speech. See IV:479–512 ("Speech on the Oregon Bill," 27 June 1848).

11. Calhoun claimed that protective tariffs were unconstitutional because the Constitution only refers to tariffs in terms of raising revenue. Merrill Peterson emphasizes the economic dimensions of the crisis. See Merrill D. Peterson, *Olive Branch and Sword—The Compromise of 1833* (Baton Rouge: Louisiana State University Press, 1984), esp., pp. 14, 88, 124.

12. See Ellis, *The Union at Risk,* pp. 93–94, 167–74. Ellis shows how Jackson, in particular, insisted on the passage of the Force Bill before he would support any compromise tariff. Relative to the economic interpretation of the controversy, it is also significant that Calhoun's speeches evidence no anticommerce bias. To him, it was always a question of how to balance an existing matrix of interests, a matrix in which the interests of South Carolina (and the other Southern states) happened to be largely agrarian. But rather than idyllize an agrarian way of life, as some Southerners did, Calhoun aggressively attempted to broaden his state's economic interests. See Theodore R. Marmor, "Anti-Industrialism and the Old South: The Agrarian Perspective of John C. Calhoun," *Comparative Studies in Society and History* 9 (1967): 377–406.

13. Cf. *Debates* IX:383 (Clayton).

14. According to Calhoun, the Constitution is not solely based on the concurrent-majority principle in that the federal government operates internally (in passing laws) on the absolute-majority principle. The amending power, furthermore, only imperfectly incorporates the concurrent-majority principle because three-fourths of the states need not include a majority of the minority on that particular issue. (Obviously, the theory works better if there is a large, permanent minority of states, which Calhoun increasingly thought was the case in America.) Just as his reading of American history received more extended treatment in the *Discourse,* he elaborated his pure theory in another posthumously published work, *A Disquisition on Government* (I:1–107).

15. Misinterpretations of Calhoun's political thought frequently are artifacts of ignoring this crucial level-difference. His thought is not only much less pluralistic on the state than on the federal level but much less undemocratic and *laissez-faire.* For example, Hartz's interpretation of Calhoun as being torn between organicism and constitutionalism is persuasive, but any inconsistency on Calhoun's part disappears when we distinguish between his organic view of the states and his strong belief in a constitutional union. See Hartz, *The Liberal Tradition in America,* pp. 158–66.

16. It is not the last word in two, rather disparate, ways. First, other narratives of the nullification crisis—constitutional, economic, sectional, even narrowly political ones—cannot simply be dismissed. My claim is "merely" that those factors, either singly or together, cannot adequately explain the crisis. Second, as already suggested, Calhoun was not strictly a federal republican. This complication will be discussed further below.

17. See *Debates* IX:255–56 (Wilkins); 328 (Theodore Frelinghuysen, D-New Jersey); 382 (Clayton); 494, 501 (William C. Rives, D-Virginia).

18. The support of the South Carolina congressional delegation, including Calhoun, for the Tariff of 1816 became an issue because that tariff was arguably also a protective one. It, thus, was alleged that the state was, at a minimum, guilty of inconsistency in now officially proclaiming all protective tariffs unconstitutional. See *Debates* IX:252–53 (Wilkins). Calhoun's retort was multifaceted. The central point, though, was his contention that the earlier tariff was only incidentally a protective one. See Calhoun II:204–12. Webster countered that if the Tariff of 1816 was not essentially a protective one, then neither was the Tariff of 1832 (see Webster VI:229, 235). By and large, however, both senators skirted the issue of the other's personal consistency on tariff policies. (Prior to the 1830s, Webster had been an advocate of free trade.)

19. See *Debates* IX:248–51 (Wilkins); 331 (Frelinghuysen); 414–15 (George Dallas, D-Pennsylvania).

20. Cf. *Debates* IX:259 (Wilkins). Following Calhoun's lead, the South Carolina electors "wasted" their ballots on Virginia Governor John Floyd rather than vote to re-elect Jackson. Calhoun's own anti-party bias has been well documented. Esp., see William W. Freehling, "Spoilsmen and Interests in the Thought and Career of John C. Calhoun," *Journal of American History* 52 (1965): 25–42.

21. See *Debates* IX:330–31 (Frelinghuysen); 399–400 (Clayton); 517 (Rives); 664 (Felix Grundy, D-Tennessee).

22. "Liberty and union" was, of course, a phrase Webster made famous during his 1830 debate with then-Senator Robert Hayne of South Carolina, something of a preview of the nullification debate. See Webster VI:75 ("Second Reply to Hayne," 26 January 1830). At the Jefferson Day dinner in Washington later that same year, Jackson proposed a toast which signaled his opposition to nullification: "Our Union—it must be preserved." Calhoun coolly responded: "The Union. Next to our liberties, most dear." See Freehling, *Prelude to Civil War,* p. 192.

23. For a thoughtful discussion of the different ideas of the union current during the Jacksonian era, see Major L. Wilson, "The Concept of Time and the Political Dialogue in the United States, 1828–1848," *American Quarterly* 19 (1967): 619–44.

24. Also, see Calhoun II: 201.

25. "Consistency of conduct" refers to Calhoun's alleged inconsistency in supporting the Tariff of 1816 but not that of 1832. "Purity of motives" refers to the accusation that his frustrated presidential ambitions have put him at the head of the nullification movement in South Carolina. Calhoun parries the latter charge by observing that he was at the height of his national influence in 1831 when he first publicly advocated nullification. He was Vice-President of the United States, with reasonable prospects of succeeding Jackson as president (see II:216–17). It was Jackson who had promulgated the charge in his "Proclamation to the People of South Carolina" of 10 December 1832. See James D. Richardson, *A Compilation of the Messages and Papers of the Presidents, 1789–1897* (Washington: Government Printing Office, 1897), p. 656. (Hereafter cited as

Jackson, "Proclamation," with appropriate page references.) Richard Latner's narrative focuses on Jackson's conspiratorial view of the crisis, interpreting that view as a traditional republican one. See Richard B. Latner, "The Nullification Crisis and Republican Subversion," *Journal of Southern History* 43 (1977): 19–38.

26. Cf. *Debates* IX:398–99 (Clayton).

27. On the possibility of submerged loyalties, see David M. Potter, "The Historian's Use of History and Vice Versa," *American Historical Review* 67 (1962): 931–32.

28. In part, the point is that Calhoun was not really the arch-nullifier, notwithstanding Jackson's claims to the contrary. He actually was a moderating influence on the "hotheads," a younger generation of South Carolina politicians who did not share Calhoun's strong attachment to the union and who, in the future, would move their state ever closer to secession. See Freehling, *Prelude to Civil War,* pp. 157–58, 221–226, 291–92.

29. Cf. *Debates* IX:247–49 (Wilkins); 332 (Frelinghuysen); Jackson, "Proclamation," pp. 654–56.

30. As noted in earlier chapters, both the Federal Farmer and Publius were more critical of Roman institutions. This contrast might suggest that Calhoun was more classically oriented than they were. I would argue that it merely indicates he was more willing to indiscriminately use historical examples which seemed to support his theory. His sympathetic treatment of the Polish *liberium* veto also betrays this tendency (see II:305–6).

31. Cf. *Debates* IX:247–49 (Wilkins); 356–57 (Holmes); 494, 502–3 (Rives). According to Calhoun, this objection is all the more groundless to the extent that the Constitution incorporates the absolute-majority principle in the internal operations of the federal government (see II:255).

32. Calhoun frequently invokes this cycle (also, see II:245, 260, 306).

33. Cf. Calhoun II:238; Publius, pp. 100–101, paper 14. Even the Federal Farmer stressed the national features of the American exemplar (see p. 281, letter IX).

34. While a number of scholars have interpreted Calhoun's concurrent-majority theory as a critique of Publius's extended-republic argument, they have fastened on his different solution, not his different problem. See Ford, "Republican Ideology in a Slave Society," pp. 411–12; Lerner, "Calhoun's New Science of Politics," pp. 924–26; Maier, "The Road Not Taken," pp. 9–10, 16–17.

35. Scholars who embrace the pluralist interpretation of Publius's extended-republic argument are fond of interpreting Calhoun's concurrent-majority theory in the same manner. See Darryl Baskin, "The Pluralist Vision of John C. Calhoun," *Polity* 2 (1969): 58–60; Bourke, "Pluralist Reading of James Madison's Tenth *Federalist,*" p. 292; Charles M. Wiltse, "Calhoun and the Modern State," *Virginia Quarterly Review* 13 (1937): 400.

36. Implicitly, then, Calhoun confronts Publius's "states too large" query of paper nine.

37. Cf. Steinberger, "Calhoun's Concept of the Public Interest," pp. 416–24.

Chapter Six

1. Rather than directly responding to Calhoun's first nullification speech, Webster gave a set speech that primarily attacked the resolutions Calhoun had introduced in the Senate spelling out the doctrines upon which South Carolina had acted. Calhoun, by the way, was one of the few senators who did not appeal to the authority of *The Federalist Papers* during the course of the debate. Cf. Webster VI:208–9, 217; *Debates* IX:295–99 (Bibb); 337 (Brown); 496, 498–99, 504, 509 (Rives); 622, 629, 633 (Poindexter). For Webster's early Federalism, see Maurice G. Baxter, *One and Inseparable: Daniel Webster and the Union* (Cambridge, MA: Harvard University Press, 1984), pp. 17–67.

2. In assessing the relative strength of their precedents, we, of course, should correct for hindsight. While the Constitution today "is" a Websterian one, that was much less true in 1833. Calhoun's precedents—principally, the Virginia and Kentucky resolutions—were powerful ones among his Senate colleagues, which was why they spent so much time debating the question of whether they really were precedents. Cf. Calhoun II:237, 277, 292, 298–99; *Debates* IX:299, 302 (Bibb); 339 (Brown); 354 (Holmes); 369, 371 (Tyler); 380–81 (Clayton); 439–43 (Miller); 499–500, 509–13 (Rives), 631–33, 638–39 (Poindexter). The academic debate is equally inconclusive. For example, cf. Harry V. Jaffa, "Partly Federal, Partly National," in *The Conditions of Freedom* (Baltimore: Johns Hopkins Press, 1975), pp. 175–77, 180–82; Maier, "The Road Not Taken," p. 13. For Webster's enormous influence as a Supreme Court lawyer, see Baxter, *One and Inseparable*, pp. 164–78.

3. In sum, Webster was not just a Publius clone, as Hofstadter argues in *The American Political Tradition*, pp. 86–87.

4. See *Debates* IX:351 (Holmes); 427–28 (Dallas); 496–97, 500 (Rives); 668 (Grundy); 681 (Thomas Ewing, NR-Ohio).

5. Ellis's narrative of the crisis emphasizes the bifurcated nature of the response to South Carolina's nullification ordinances, with the nationalistic response coloring but not controlling the constitutional one. See Ellis, *The Union at Risk*, pp. ix, 12, 88, 183. In this respect, Jackson's proclamation was paradigmatic.

6. For the standard view of Webster's speech and contemporaneous actions, see Norman D. Brown, *Daniel Webster and the Politics of Availability* (Athens: University of Georgia Press, 1969), pp. v–vi, 15–66. In assessing this view of Webster's speech, we should recall that Jackson's proclamation horrified many of his own party leaders (in particular, Martin Van Buren) for its "ultra"-nationalism. Yet, it was not as "ultra" as Webster's speech. In the background, moreover, there was Jackson's highly contentious veto of the bill rechartering the second Bank of the United States. Webster not only had strongly supported the recharter bill, but he claimed that the president's veto message looked askance to nullification. See Webster II:121–23 ("National Republican Convention at Worcester," 12 October 1832). This is not to deny that Jackson's proclamation and Webster's speech were close in spirit and that the crisis temporarily eased the friction, both political and

personal, between them. Crises, however, do have that tendency without necessarily signaling an attempt to forge a political alliance.

7. See Current, *Daniel Webster and the Rise of National Conservatism,* pp. 64–69.

8. For example, see Webster II:148–49 ("Reception at Pittsburg[h]," 8 July 1833); II:203 ("Reception at New York," 15 March 1837); III:177–78 ("Convention at Andover," 9 November 1843); IV:159–60 ("Speech at Faneuil Hall," 24 October 1848). The nonrepublican or narrowly liberal view of Webster (and the Whig Party) concentrates on his defense of protective tariffs and commercial expansion. See Diggins, *The Lost Soul of American Politics,* pp. 116, 136. It, again, is beyond my present purposes to fully address this view. Briefly, I would say that the Whigs, like the Federalists before them, defended commercial expansion on identifiably republican grounds. In his seminal study of the Whig Party, Howe argues that it combined procommerce and republican values. See Daniel Walker Howe, *The Political Culture of the American Whigs* (Chicago: University of Chicago Press, 1979), pp. 9, 101, 299. For an ingenious analysis of how Webster manipulated traditional agrarian (republican?) values to vindicate the virtues of the new industrial classes, see Clarence Mondale, "Daniel Webster and Technology," *American Quarterly* 14 (1962): 37–47.

9. Indeed, both state an intention not to discuss the tariff. See Webster VI:228; Calhoun II:207.

10. The previously cited articles by Jaffa ("Partly Federal, Partly National") and Major Wilson ("Liberty and Union") provide good examples of such narratives.

11. Robert Dalzell's account of the last decade of Webster's life remains the best work on Webster, in part because it nicely captures the complex relationship which existed between his nationalism, his presidential ambitions, and the economic interests of his constituents. See Robert F. Dalzell, *Daniel Webster and the Trial of American Nationalism, 1843–1852* (Boston: Houghton Mifflin, 1973). I am not aware of an explicitly republican interpretation of Webster, although, as already noted, Howe treats republicanism as one of the major strands in Whig political culture.

12. Cf. Calhoun II:200–204, 294–99. Webster cites Federalists Madison and Pinckney as well as Anti-Federalist Luther Martin to support his interpretation of the Constitution (see VI:216–17). Calhoun counters that Martin's public statements on the Constitution cannot be trusted due to his intense opposition to the plan (see II:230).

13. Cf. Publius, p. 476, paper 80. Calhoun contends that the convention delegates did not contemplate judicial review, at least not with respect to state laws (see II:203).

14. Cf. Publius, p. 152, paper 22. The quoted passage, however, only shows that Publius rejected the idea of founding the new government on a compact between the state governments, which was how the Articles of Confederation were ratified. (Even Calhoun repudiated that idea. For him, one of the critical departures of the Constitution was that it was ratified by the people of the states [see II:289–91].) The passage does not show that Publius rejected the compact view *in toto.* Elsewhere, he seems to subscribe to such a view (see pp. 37, paper 2; 204,

paper 34; 243–44, paper 39; 324–25, paper 51; 524–25, paper 85; cf. Federal Farmer, pp. 228, letter I; 246, letter IV; 331, letter XVII). Most scholars agree that the founders viewed the Constitution as a compact of some kind. See Andrew C. McLaughlin, "Social Compact and Constitutional Construction," *American Historical Review* 5 (1900): 472–82; Thad W. Tate, "The Social Contract in America, 1774–1787," *William and Mary Quarterly,* 3d series, 22 (1965): 375–91; Charles M. Wiltse, "From Compact to National State in American Political Thought," in *Essays in Political Theory, Presented to George H. Sabine,* edited by Milton R. Konvitz and Arthur E. Murphy (Ithaca, NY: Cornell University Press, 1948), p. 155.

15. Predictably, Calhoun criticized Rives's speech on the opposite grounds (see II:236–37). Both Webster and Calhoun focused on Rives because he occupied the ideological center of the debate and, not coincidentally, because he was close to Madison. See Ellis, *The Union at Risk,* p. 183. The former president was the reigning member of the Virginia states-rights school, the "very distinguished school" to which Webster refers. Madison made a number of public and semipublic pronouncements denouncing nullification as repugnant to the letter and spirit of the Constitution. Esp., see James Madison, "Letter to the Editor," *North American Review* 31 (1830): 537–46. Rives cited this letter in his speech (see *Debates* IX:496–97).

16. The difference between Calhoun and Rives appears to be the difference between Calhoun and the Federal Farmer. The key is whether individual states can suspend federal laws before obtaining the support of at least a majority of the other states. Madison denied that the Virginia Resolutions constituted a precedent for South Carolina's nullification ordinances on similar grounds. While both states had appealed to the amending process, South Carolina, by unilaterally nullifying the tariffs, had reversed the logic of that process. See Madison, "Letter to the Editor," pp. 543–45. Also, James Madison, "Notes on Nullification," in *The Mind of the Founder: Sources of the Political Thought of James Madison,* edited by Marvin Meyers (Indianapolis: Bobbs-Merrill, 1973), pp. 547–58.

17. On the eve of the war, Lincoln promulgated this view of its causes. See Roy P. Basler (ed.), *Abraham Lincoln: His Speeches and Writings* (New York: Da Capo, 1946), pp. 602–3 ("Message to Congress," 4 July 1861). [Hereafter cited as Basler, with appropriate page and speech references.]

18. Webster, consequently, disproves Jaffa's claim that all the major figures in the debate agreed on a compact philosophy. See Jaffa, "Partly Federal, Partly National," p. 166. Cf. Nagel, *One Nation Indivisible,* p. 39; Wilson, "Liberty and Union," pp. 337–40; Wiltse, "From Compact to National State," p. 175.

19. Calhoun uses this language, from the Massachusetts ratifying convention no less, against Webster (see II:276–77).

20. For different assessments of the historical evidence supporting Webster's claim of a "pre-existing" union, cf. Curtis P. Nettels, "The Origins of the Union and of the States," *Proceedings of the Massachusetts Historical Society* 72 (1957–60): 68–83; Kenneth M. Stampp, "The Concept of a Perpetual Union," *Journal of American History* 65 (1978): 6–9. Calhoun naturally denies the claim (see II:281–82). Jackson, though, asserts it in even stronger language than Webster does (see Jackson, "Proclamation," pp. 643, 650).

21. Again, cf. Jackson, "Proclamation," pp. 648–49. For secondary discussion, see Stampp, "Concept of a Perpetual Union," pp. 28–32.

22. Cf. *Debates* IX:256–57 (Wilkins); 274, 304–5 (Bibb); 314, 323–24 (Frelinghuysen); 364–65 (Tyler); 383–86 (Clayton); 422 (Dallas); 495–96 (Rives); 668, 673 (Grundy); 679–83 (Ewing); Jackson, "Proclamation," pp. 643, 648; Madison, "Letter to the Editor," pp. 537–38.

23. Cf. Calhoun II:282–83.

24. See *Debates* IX:257 (Wilkins); 351–52 (Holmes); 423 (Dallas); 504 (Rives); 668 (Grundy); 682–83 (Ewing); Jackson, "Proclamation," 649–50; Madison, "Letter to the Editor," p. 537. It, of course, is a subject of much debate whether Madison had substantially changed his position in this and other respects since his contributions to *The Federalist Papers.* Cf. Jaffa, "Partly Federal, Partly National," pp. 163, 173; Jean Yarbrough, "Madison and Modern Federalism," in *How Federal Is the Constitution,* edited by Robert A. Goldwin and William A. Schambra (Washington: American Enterprise Institute, 1987), pp. 84–108.

25. There seems to be a growing appreciation of Webster as a literary figure—part of the same American Romantic movement as Emerson—and not just as a legal or political figure. Esp., see Dalzell, *Daniel Webster and the Trial of American Nationalism,* pp. xiii–xiv. Even though they provide overly legalistic glosses on this particular speech, Paul Erickson and Robert Ferguson generally stress Webster's literary aspirations. See Paul D. Erickson, *A Poetry of Events: Daniel Webster's Rhetoric of the Constitution and Union* (New York: New York University Press, 1986), pp. xii–xiii, 41, 106, 128; Robert A. Ferguson, *Law and Letters in American Culture* (Cambridge: Cambridge University Press, 1984), pp. 208–9, 228. For the older view, see Hartz, *The Liberal Tradition in America,* pp. 109, 165.

26. In observing that Webster does not really discuss a common interest in this speech, I, thus, am not denying that it contains a number of assertions of a common interest. Yet, he still does not discuss how, in general, there is a common interest nor how, in particular, it is served by a protective tariff. As noted above, these are topics he does discuss in many of his other speeches.

27. Webster, however, is not supremely confident. His peroration goes on to sketch a scenario of an escalating crisis. In the introduction of the speech, he recalls how he was much less confident in 1830, at the time of the Hayne debate, and modestly admits that his efforts then probably have contributed to the improved state of public opinion (see VI:183, 237–38).

28. Although, as we saw in chapter 4, Publius did not neglect the sentimental union, it clearly is much more central to Webster's political argument. See Paul C. Nagel, "Democratic Thought and the Symbol of Union: Early Phases," *Mississippi Quarterly* 12 (1959): 58–59. For the growth of popular nationalism in the intervening period, see Kohn, *The Idea of Nationalism,* pp. 287, 322.

29. Webster, then, emphasizes that the crucial difference between the Articles of Confederation and the Constitution is federal powers in the sense of modes, not objects. Webster cites two founders, Samuel Johnston and Oliver Ellsworth, for confirmation. He also contends that even the opponents of the Constitution agreed that it meant a fundamental change in the mode of exer-

cising federal powers (see VI:207–8). As we have seen, both Publius and the Federal Farmer stressed, and favored to different extents, this change. Calhoun, not surprisingly, does not distinguish the two systems in this way.

30. Frelinghuysen decries state pride as the cause of the nation's periodic crises (see *Debates* IX:314). Also, see Jackson, "Proclamation," pp. 648, 652.

31. This urgency certainly was palpable in Webster's controversial "Seventh of March" (1850) speech at another critical moment in North-South relations (see X:57–58, 92–93).

32. In this context, we should recall two additional factors mentioned in the course of the preceding discussion. First, Webster does not actually surrender the traditional republican assumption of social homogeneity. He merely suggests its limits. Second, there is at least one sense in which national republicanism now might be more socially realistic. This is the extent to which patriotism is Americans' uppermost sentiment.

33. While he does not identify this change as part of the evolution of an American liberal tradition, Nagel argues that Americans were much more accepting of economic and cultural diversity in the 1830s than they had been in the 1780s. (After all, there was more of it!) See Nagel, *One Nation Indivisible,* pp. 90, 111–12, 198. Howe, in contrast, claims that the two major parties divided in their attitudes toward diversity. The Whigs were more accepting of economic diversity; the Jacksonians, of cultural diversity. See Howe, *The Political Culture of the American Whigs,* pp. 20, 35.

34. Webster does not even explicitly endorse the Force Bill, though his speech does lay the constitutional and theoretical foundations for such a bill.

35. Ultimately, the Federal Farmer, Publius, Calhoun, and Webster all portray their opponents in conspiratorial terms. Recent scholarship has reopened the question of whether American statesmen had discarded their extremely negative attitudes toward political parties by the time of the emergence of the second-party system in the mid-1830s. A subquestion is whether the Whigs lagged far behind the Jacksonians in this respect. Cf. Ashworth, *'Agrarians' and 'Aristocrats,'* pp. 205–18; Ronald P. Formisano, "Deferential-Participant Politics: The Early Republic's Political Culture, 1789–1840," *American Political Science Review* 68 (1974): 473–87; Perry Goldman, "Political Virtue in the Age of Jackson," *Political Science Quarterly* 87 (1972): 46–62; Richard Hofstadter, *The Idea of a Party System: The Rise of Legitimate Opposition in the United States, 1780–1840* (Berkeley: University of California Press, 1969), chap. 6. For Webster's own anti-party bias, see Howe, *The Political Culture of the American Whigs,* pp. 222–23.

36. Calhoun seems bemused by Webster's pugilistic metaphor (see II:307).

37. For an insightful analysis of Webster's ethical appeals in the Hayne debate, see Wayne Fields, "The Reply to Hayne: Daniel Webster and the Rhetoric of Stewardship," *Political Theory* 11 (1983): 5–28.

38. Also, see Madison, "Letter to the Editor," p. 543.

39. Webster, nonetheless, does also depict the conditions under the Articles of Confederation and, hypothetically, nullification as anarchic (see pp. 194, 212).

40. Webster, therefore, anticipates later historians in overlooking the level-difference in Calhoun's position. In a draft of a letter he never sent, Madison acknowledges that Calhoun's objection is not to majority rule *per se* but to majority rule as it operates on a large, national scale. Madison then launches into a restatement of his extended-republic argument of *Federalist* 10. See James Madison, "Majority Governments (1833)," in *The Mind of the Founder,* pp. 521–30.

41. Webster duly notes how the Constitution limits majority rule both in the elective and legislative processes (see VI:220). However, these "internal" limitations on majority rule are at issue only in the sense that he believes they are sufficient on the federal level and Calhoun does not.

42. Many of the other senators participated in this struggle. See *Debates* IX:312 (Bibb); 332 (Frelinghuysen); 376 (Tyler); 433 (Miller); 490 (Moore); 507, 517 (Rives); 642, 652 (Poindexter).

43. Webster was certainly open to this criticism in regard to the slavery issue. Even if he thought that disunion would only further ensconce the institution in the Southern states, he still was not overly imaginative in presenting political options other than the union, as is, or disunion. See Ferguson, *Law & Letters,* p. 227.

44. Webster is invariably credited with being one of the, if not *the,* major champion(s) of the American mission during the pre–Civil War period. Esp., see Nagel, *One Nation Indivisible,* pp. 157–58, 161–63.

45. See Calhoun, II:257, 301. Also, *Debates* IX:292 (Bibb); Madison, "Notes on Nullification," p. 574.

46. Here, Webster is stressing the "sacred trust" aspect of mission; that is, the obligations succeeding generations of Americans owe to the founders to preserve their Constitution and union. See Nagel, *This Sacred Trust,* pp. 88–90, 158–60. The standard interpretation of Webster as a political conservative does have considerable weight behind it. The problem with this interpretation is that it proves too much. It, for instance, does not distinguish a Webster, when he is trying to preserve liberal institutions, from a Calhoun, when he is trying to preserve nonliberal institutions. Cf. Merrill D. Peterson, *The Great Triumvirate: Webster, Clay, and Calhoun* (New York: Oxford University Press, 1987), pp. 395–400, 409–11.

47. The list also exhibits the disconcerting mix of public and private goods which seems definitive of liberal thought. In this mix, Webster does appear to emphasize the public portion more than a pluralist would. That clearly is the case on liberty, where it is the public liberty associated with republican governments, not the private liberty associated with the absence of rights violated and wrongs unredressed, which defines the American mission for him (see VI:237).

48. See Nagel, *One Nation Indivisible,* pp. 20, 254.

Chapter Seven

1. In part, the point is that Douglas, in 1858, cannot convincingly portray either the union or the states in traditional republican terms. This is not to deny that Calhoun and the Federal Farmer had to engage in a certain amount of hy-

perbole to portray the states as traditional republics. Still, they were able to make their portrayals somewhat convincing; Douglas did not even think of trying.

2. Again, we should recall Publius's claim that the Anti-Federalist position resulted in a similar absurdity.

3. See Damon Wells, *Stephen Douglas: The Last Years* (Austin: University of Texas Press, 1971), pp. 67–73. The division of the Kansas-Nebraska territory to give slavery a chance in Kansas exposed the doctrine's ambiguity on locus. Similarly, the doctrine's ambiguity on timing could at least temporarily paste over the differences between northern Democrats (who favored an early decision to exclude slavery) and southern Democrats (who favored a late decision to, again, give slavery a chance). Although both sides agreed that when a territory made a decision on slavery would have an important effect on the outcome of that decision, they did not agree that the decision itself would have any effect, legally because of the *Dred Scott* decision and practically because of the thesis that slavery had already reached the natural limits of its growth in the United States. Douglas, of course, attempted to address the legal question through his (in)famous "Freeport Doctrine" and tended to subscribe to the "natural limits on slavery" thesis. The latter view made Douglas's position, in effect, the same as the Republicans' free-soil position. However, as we will see over the course of the next two chapters, the difference in principle remained highly significant. Cf. Harry V. Jaffa, *Crisis of the House Divided: An Interpretation of the Issues of the Lincoln Douglas Debates* (Chicago: University of Chicago Press, 1982), pp. 114–26.

4. For an overview of prior interpretations of Douglas, see Stephen B. Oates, "Little Giant Reconsidered," *Reviews in American History* 1 (1973): 534–41. Oates focuses on what remains the most thorough biography: Robert W. Johannsen, *Stephen A. Douglas* (New York: Oxford University Press, 1973). Johannsen emphasizes Douglas's nationalism. Esp., see pp. vii–viii, 139, 264, 322, 772–73, 866.

5. Douglas probably owes his doctrine to Lewis Cass, a longtime party colleague and senator from Michigan. Cass first suggested the doctrine in a letter to a political supporter during his unsuccessful 1848 presidential campaign. See Johannsen, *Stephen A. Douglas*, pp. 227–28.

6. The two opening speeches are included in Johannsen's edition of the debates. See Robert W. Johannsen (ed.), *The Lincoln-Douglas Debates* (New York: Oxford, 1965). [Hereafter cited as Johannsen, with appropriate references to page and speaker.] The fact that Lincoln was officially nominated for the Senate at a state party convention was unprecedented. Previously, the practice had been for the state legislature to elect the "best man." (By this time, he was invariably chosen by the majority-party caucus from among the state's most prominent party members.) Douglas claimed that the Illinois Republicans took this unusual step in order to pacify Lincoln for what happened in 1856 when, according to Douglas, Lyman Trumbull had reneged on an agreement to support Lincoln's election to the Senate in order to secure the seat for himself (see Johannsen, pp. 43, 100, 118–24, 155–56, 189–93 [Douglas]). The more likely reason was to prevent Douglas from emerging as a fusion candidate based on his opposition to Kansas's proslavery Lecompton constitution, which saw him side

with the Republicans in Congress against the Democratic administration. During the course of the ensuing campaign, each candidate delivered more than a hundred speeches, sometimes in tandem. Indeed, Lincoln's initial strategy was to follow his more famous opponent around the state to take advantage of Douglas's crowds. This strategy proved unsatisfactory to both candidates and led them to formally schedule seven joint appearances. For more on the background of the debates, see Donald E. Fehrenbacher, *Prelude to Greatness: Lincoln in the 1850's* (Stanford, CA: Stanford University Press, 1962), chaps. 3–5.

7. The Federal Farmer had highlighted slavery as *an* example of interstate diversity.

8. Also, see Johannsen, pp. 44–45, 126–27, 288 [Douglas].

9. Douglas contemplates the possibility of acquiring Cuba, parts of Canada, and even more of Mexico. For the centrality of national expansion and manifest destiny to Douglas's political vision, see Robert W. Johannsen, "Stephen A. Douglas and the American Mission," in *The Frontier, the Union and Stephen A. Douglas* (Urbana: University of Illinois Press, 1989), pp. 77–98. Of Douglas, Lincoln, Calhoun, and Webster, Douglas was the only one to unequivocally support expansion. The other three did not support expansion when, for instance, they thought it might exacerbate the slavery issue. See Johannsen, pp. 77, 234–36 [Lincoln]; Calhoun IV:323–24 ("On 'The Three-Million Bill,'" 9 February 1847); Webster X:12, 20–21, 32 ("Objects of the Mexican War," 23 March 1848). They, however, were not opposed to expansion on general principles. Cf. Jaffa, *Crisis of the House Divided,* pp. 67–69, 77–81, 99; Major Wilson, *Space, Time, and Freedom,* pp. 114–16.

10. Also, see Johannsen, pp. 126, 197, 276 [Douglas].

11. While earlier generations had agreed that Congress has the power to regulate slavery in the territories, they had disagreed over whether it was expedient for Congress to exercise that power. Only in the 1850s, particularly after the *Dred Scott* decision, did the Southern position become that Congress should (must) adopt legislation favorable to the expansion of slavery into the territories (such as a territorial slave code), just as the Northern position became that Congress should absolutely prohibit slavery from entering the territories. See Arthur Bestor, "The American Civil War as a Constitutional Crisis," *American Historical Review* 69 (1964): 346–47, 351.

12. See Stephen A. Douglas, "The Dividing Line between Federal and Local Authority," in *In the Name of the People: Speeches and Writings of Lincoln and Douglas in the Campaign of 1859,* edited by Harry V. Jaffa and Robert W. Johannsen (Columbus: Ohio State University Press, 1959), pp. 58–125. Douglas had long been the most powerful congressional advocate of territorial rights, and not just on the slavery issue. He claimed that traditionally Congress had unfairly treated territories as colonies, not as prospective states. Also see Johannsen, p. 328 [Douglas]; Robert W. Johannsen, "Stephen A. Douglas, *Harper's Magazine,* and Popular Sovereignty," in *The Frontier, the Union and Stephen A. Douglas,* pp. 120–39.

13. Actually, no state had positively abolished slavery by 1787–88, though four states (New Hampshire, Massachusetts, Connecticut, and Pennsylvania) had moved to abolish slavery either through judicial or legislative action. See

Arthur Zilversmit, *The First Emancipation: The Abolition of Slavery in the North* (Chicago: University of Chicago Press, 1967), chap. 5. Yet, it is extremely doubtful that any state, North or South, would have then sought to make slavery a permanent, national institution.

14. Douglas believes the new free-state majority also accounts for the rise of the Republican Party as a purely sectional party (see Johannsen, pp. 211, 300 [Douglas]).

15. Douglas's use of the Compromise of 1850 is part of a concerted effort to separate Lincoln from Clay's Whig legacy, a legacy which, as we shall see in the next chapter, Lincoln studiously cultivates.

16. The Kansas-Nebraska Act recalls one of the critical pieces of legislation Douglas skips over in his historical narrative—the Missouri Compromise of 1820. This "oversight" is hardly surprising, given that the Missouri Compromise was antithetical to popular-sovereignty doctrine. Douglas, of course, insists that the Compromise of 1850 implicitly repealed the Missouri Compromise; the Kansas-Nebraska Act only made that repeal explicit. For an extended discussion of the intricacies of this legislative history, see Jaffa, *Crisis of the House Divided,* chaps. 5–8.

17. Douglas's "Freeport Doctrine" was part of these efforts. For further discussion, see Fehrenbacher, *Prelude to Greatness,* chap. 6.

18. Douglas goes on to charge the Buchananites and Republicans with forming an unholy alliance to unseat him (see Johannsen, pp. 35–36, 107, 190, 209–11, 294–95 [Douglas]). Lincoln denies the charge but there is some historical evidence to support it. Cf. Johannsen, pp. 226, 278–79, 300–301 [Lincoln]; Fehrenbacher, *Prelude to Greatness,* pp. 113–14.

19. This injunction extends to the impropriety of the federal government throwing its weight on the side of the institutions of one or more states in cases of conflict (see Johannsen, pp. 293–94 [Douglas]).

20. Chapter 8 will largely be devoted to examining just how and why Lincoln resists this extension of the noninterference principle.

21. Because of its patterns of settlement, Illinois was a microcosm of the United States on the slavery issue, as on most other issues. See Christopher N. Breiseth, "Lincoln, Douglas, and Springfield in the 1858 Campaign," in *The Public and the Private Lincoln: Contemporary Perspectives,* edited by Cullom Davis, Charles B. Strozier, Rebecca Monroe Veach, and Geoffrey C. Ward (Carbondale: Southern Illinois University Press, 1979), p. 101.

22. Also, see Johannsen, pp. 238, 288 [Douglas].

23. Also, see Johannsen, pp. 212–15, 237 [Douglas].

24. For different assessments of the validity of the charge, cf. Hofstadter, *The American Political Tradition,* p. 148; Jaffa, *Crisis of the House Divided,* pp. 366–68.

25. In both cases, Douglas presumes that Lincoln might do something untoward. He, consequently, accuses Lincoln of appealing to mob rule to overturn the *Dred Scott* decision and of trying to literally starve slavery out of existence in the South by denying it the room to expand (see Johannsen, pp. 32, 242–44, 266–69, 327–28 [Douglas]). Both accusations, of course, caricature Lincoln's position, as we will see in chapter 8.

26. This argument assumes that Douglas was personally opposed to slavery but that he did not publicly state that opposition because it would have violated his own understanding of democracy to do so. I do not mean to deny the other, perhaps equally important, reasons for his silence. He also felt that taking an antislavery stand would hurt his political career, dishonor his commitment to an intersectional Democratic Party, and endanger the union. See Wells, *Stephen Douglas*, pp. 21, 107–9.

27. In this respect, Douglas's views are reminiscent of Calhoun's idealized concurrent-majority process. The comparison to Publius is also apt in the sense that Douglas believed fluctuating majorities and minorities would exist in modern democracies, thereby reducing the possibility of majority imposition or tyranny. Finally, Douglas, now paralleling the Federal Farmer and Calhoun more than Publius and Webster, stressed the role of federalism in reducing the possibility of such an imposition.

28. The Federal Farmer, Publius, Calhoun, and Webster primarily expressed the then-dominant republican strain. Publius is clearly the least like Douglas. Indeed, Douglas's political psychology seems diametrically opposed to Publius's, as in the oft-quoted passage from *Federalist Paper* 72 on "the love of fame" (see p. 437).

29. The Douglas-Publius comparison is, again, apt. They both anticipated twentieth-century views on social diversity, but neither would have been able to fully accept those views.

30. Jaffa interprets Douglas as a majoritarian democrat. He, however, does not claim that Douglas's majoritarianism made him illiberal. He, rather, argues that Lincoln transcended liberalism insofar as he rebutted Douglas's majoritarianism on substantive grounds, an argument which would make Douglas liberal. See Jaffa, *Crisis of the House Divided,* chaps. 13–17. Nonetheless, according to more standard views of liberalism as being incompatible with strict majority rule, Jaffa's interpretation of Douglas would make him illiberal and Lincoln liberal.

31. At the time, there were three relevant combinations on the question of the limits on majority rule: majority rule versus the rights of individual citizens (white or black); majority rule versus the rights of white noncitizens; majority rule versus the rights of black noncitizens (free or slave). Many Americans had separate answers on each combination.

32. It is a measure of Douglas's consistency that he vows not to complain about the policies of the Northern states that do allow blacks to become full citizens. Only the five New England states of Maine, New Hampshire, Vermont, Massachusetts, and Rhode Island granted free blacks the same voting rights as whites. See Leon F. Litwack, *North of Slavery: The Negro in the Free States, 1790–1860* (Chicago: University of Chicago Press, 1961), p. 91.With pejorative intent, Lincoln suggests that *Dred Scott* decided the Constitution does preclude black citizenship. In Douglas's hands, this suggestion becomes a statement to the effect that Lincoln has "especially" criticized *Dred Scott* for ruling against black citizenship. Lincoln quickly renounces that statement. Cf. Johannsen, pp. 32–33, 45, 127 [Douglas]; pp. 17, 198, 301–2 [Lincoln].

33. Also, see Johannsen, pp. 33, 45–46, 196, 215–16, 299 [Douglas].

34. Again, this is not to deny that other factors were also involved in their different public stances on the slavery issue.

Chapter Eight

1. Not surprisingly, Lincoln most explicitly stated his views on liberty and union during the secession winter of 1860–61. See Basler, p. 577 ("Address in Independence Hall," 22 February 1861); pp. 582–83 ("First Inaugural Address," 4 March 1861). Webster's "Second Reply to Hayne," which contains the "liberty and union" peroration, was one of Lincoln's models for his first inaugural. See Richard N. Current, "Lincoln and Daniel Webster," *Journal of the Illinois State Historical Society* 48 (1955): 319–21. However, Lincoln, in his debates with Douglas, emphasized his debt to Clay and did not even mention Webster. The explanation is political geography. Clay, like Lincoln, was a western Whig and Clay, unlike Webster, had been popular in Illinois; especially in central Illinois, which was critical to Lincoln's prospects for election. (This is also why Douglas so industriously tried to block Lincoln's efforts to bask in Clay's legacy.) See David Zarefsky, *Lincoln, Douglas and Slavery: In the Crucible of Public Debate* (Chicago: University of Chicago Press, 1990), pp. 155–63. I, nevertheless, would argue that Lincoln was closer ideologically to Webster than he was to Clay. Unlike Lincoln and Webster, Clay's nationalism always remained essentially economic and his views on slavery hardened as he grew older. But, cf. George M. Fredrickson, "A Man but Not a Brother: Abraham Lincoln and Racial Equality," *Journal of Southern History* 41 (1975): 41–44.

2. At this point, a number of disclaimers are in order. I am not arguing that Lincoln's and Webster's definitions of liberty differed but that their definitions of union and of liberty *and* union differed. Greenstone makes a similar argument, although we disagree over how to best understand the Webster-Lincoln differences. See Greenstone, "Political Culture and American Political Development," pp. 42–47. (I will also follow Greenstone in comparing Lincoln to Douglas on liberty and union. Here, however, our disagreement is greater because he sees Lincoln and Douglas as holding different definitions of liberty and I do not.) I am also not arguing that Lincoln's definition of liberty and union was devoid of particularistic elements or, as I noted in chapter 6, that Webster's definition was devoid of universalistic ones. Nor, finally, am I arguing that Lincoln's definition of union was devoid of republican elements or that Webster's definition was devoid of pluralistic ones.

3. See Basler, pp. 322–23 ("The Repeal of the Missouri Compromise and the Propriety of its Restoration: Speech at Peoria, Illinois, in Reply to Senator Douglas," 16 October 1854). Lincoln did not quote this part of the speech during his debates with Douglas.

4. See Jaffa, *Crisis of the House Divided,* pp. 155–57, 390.

5. In part, this "gap" would have been just a function of the generational difference between Webster and Lincoln. The whole matter is obviously speculative. The only certainty is that the Webster of 1850 was not as radical on the slavery issue as the Lincoln of 1858.

6. Also, see Johannsen, pp. 316–17 [Lincoln].

7. Bestor argues that the free-soil policy favored by Lincoln and other Republicans became the fulcrum of national debate on the slavery issue because of a constitutional consensus which precluded more radical antislavery policies. See Bestor, "The American Civil War as a Constitutional Crisis," pp. 333–34, 340–45, 351–52.

8. As we have seen, both Webster and Douglas sometimes claimed that they supported policies which would, in effect, exclude slavery from all the new territories. Even if that were true, their stances did not have the same edifying effect of an explicit free-soil policy. Indeed, I will argue that for Lincoln public edification was the primary purpose of such a policy.

9. In particular, neither Lincoln nor Douglas argued for as much social coherence as Webster and Publius did or as much as Calhoun and the Federal Farmer did on the state level.

10. While it is undoubtedly true that Lincoln was more willing to risk the union for the liberty of the slaves than Douglas was, it is also true that Douglas had a higher assessment of the risk. Webster faced a parallel situation in 1850, when he argued for the compromise measures in opposition to much of the future Republican leadership. See Major L. Wilson, "Of Time and the Union: Webster and His Critics in the Crisis of 1850," *Civil War History* 14 (1968): 293–308.

11. In making this claim, I do not mean to suggest that federalism or states' rights was not a complicating factor in the Lincoln-Douglas debates; just that it was not the fundamental factor. I have already stated this argument in chapter 7 and I will develop it further in this chapter. Interestingly, Bestor contends that many Southerners, despite their states-rights rhetoric, were, in essence, taking a nationalist position on the slavery issue by insisting on federal protection for their peculiar institution. See Bestor, "State Sovereignty and Slavery," pp. 120, 165.

12. At the beginning of the campaign, Lincoln coyly portrayed his doctrine as simply a prediction, in part to deflect Douglas's charge that he was fomenting war between the states. See Basler, p. 392 ("Reply to Douglas at Chicago," 10 July 1858); p. 415 ("Reply to Douglas at Springfield," 17 July 1858). By the time of the debates, however, he had dropped all pretenses of neutrality (see Johannsen, pp. 54–55 [Lincoln]).

13. Lincoln's overture to Douglas at the end of the "House Divided" speech to join the Republican Party was, therefore, not merely a political gambit (see Johannsen, p. 21 [Lincoln]).

14. Douglas always seemed more perspicacious than Lincoln about Southern intentions. See Allan Nevins, "Stephen A. Douglas: His Weaknesses and His Greatness," *Journal of the Illinois State Historical Society* 42 (1949): 409–10. Lincoln, nonetheless, may not have overestimated the strength of American nationalism in the South so much as he underestimated its weakness in the face of a highly aroused and well-organized disunionist minority. Lincoln's 1861 claim that the majority of Southerners did not favor secession was probably an accurate one. See Basler, p. 606 ("Message to Congress," 4 July 1861). Cf. Avery O. Craven, *The Growth of Southern Nationalism, 1848–1861* (Baton Rouge: Louisiana

State University Press, 1953), pp. 399–400; Potter, "The Historian's Use of Nationalism," pp. 943–48.

15. The two most prominent revisionist interpretations of Lincoln and the Republican Party are Eric Foner, *Free Soil, Free Labor, Free Men: The Ideology of the Republican Party before the Civil War* (New York: Oxford University Press, 1970), esp., pp. 9–10, 58–65, 308–13; Michael F. Holt, *The Political Crisis of the 1850s* (New York: Norton, 1978), chap. 7. Also, see David Brion Davis, *The Slave Power Conspiracy and the Paranoid Style* (Baton Rouge: Louisiana University Press, 1969), chap. 3; Larry Gara, "Slavery and Slave Power: A Crucial Distinction," *Civil War History* 15 (1969): 5–18.

16. See Fehrenbacher, *Prelude to Greatness,* pp. 80–82; Jaffa, *Crisis of the House Divided,* pp. 277–78. Zarefsky provides a thorough—probably too thorough—discussion of Lincoln's (and Douglas's) use of conspiracy theories. See Zarefsky, *Lincoln, Douglas and Slavery,* chap. 3.

17. Lincoln, moreover, could have believed some Southerners desired the indefinite expansion and perpetuation of the institution of slavery without believing they constituted a slavocracy.

18. See David Donald, "Abraham Lincoln, Politician," in *Lincoln Reconsidered: Essays on the Civil War Era* (New York: Knopf, 1956), pp. 57–81. It is generally agreed that by the 1850s (if not sooner) most Americans held favorable attitudes toward political parties. See Hofstadter, *The Idea of a Party System,* pp. 267–71. One indication of this fact during the Lincoln Douglas debates was "the battle of the platforms." In the end, both Lincoln and Douglas uncovered local party platforms which were too radical for the other one to support yet each claimed to support his own "official" party platform (cf. Johannsen, pp. 49, 79–80, 108–9, 137–44, 227–29, 245–46, 282–84 [Lincoln]; pp. 40–41, 70, 97, 102–3, 156–57, 239–41, 257–59 [Douglas]). This battle was silly, but it does show that neither man wanted to be portrayed as disloyal to his own party.

19. Scholars have long debated the question of whether the institution of slavery was becoming increasingly unprofitable and whether its confinement to the Southern states would have been its deathknell. See Robert W. Fogel and Stanley L. Engerman, *Time on the Cross: The Economics of American Negro Slavery* (Boston: Little, Brown, 1974), chaps. 3, 5.

20. The question of whether legal restrictions were necessary to keeping slavery out of the new territories also has long been the subject of academic debate. Cf. Eugene D. Genovese, *The Political Economy of Slavery: Studies in the Economy and Society of the Slave South* (Middletown, CT: Wesleyan University Press, 1989), chap. 10; Jaffa, *Crisis of the House Divided,* chap. 18.

21. Lincoln was touting the twin policies of gradual, compensated emancipation and colonization of the freed slaves as late as 1862. He, however, was frustrated by the border states' refusal to implement such policies and soon thereafter he decided to issue the Emancipation Proclamation. See James M. McPherson, *Abraham Lincoln and the Second American Revolution* (New York: Oxford University Press, 1991), pp. 33–34.

22. See Aileen S. Kraditor, *Means and Ends in American Abolitionism: Garrison*

and His Critics on Strategy and Tactics, 1834–1850 (New York: Pantheon, 1969), pp. 26–28.

23. Esp., see Basler, pp. 493–504 ("Annual Address Before the Wisconsin State Agricultural Society," 30 September 1859). In a well-known gloss on Lincoln's economic views, Hofstadter criticizes him for completely failing to anticipate the future growth of American capitalism. See Hofstadter, *The American Political Tradition,* pp. 135–36. Even if this criticism is fair, it does not deny (nor does Hofstadter intend it to deny) the fact that Lincoln was committed to a capitalistic economic system. Also, see Norman Graebner, "The Apostle of Progress," in *The Public and the Private Lincoln,* pp. 71–85; Howe, *The Political Culture of the American Whigs,* pp. 280, 297. For Douglas's commitment to such a system, see Johannsen, "Stephen A. Douglas and the American Mission," pp. 81–82.

24. For example, see Charles B. Strozier, *Lincoln's Quest for Union: Public and Private Meanings* (Urbana: University of Illinois Press, 1987). These "ulterior-motive" accounts invariably quote Lincoln's long-time law partner, William Herndon, to the effect that his ambition was "an engine which knew no rest."

25. See Greenstone, "Political Culture and American Political Development," pp. 36–42. The complement to the Herndon quote is Lincoln's response to Douglas's charge that the Republican Party is the machination of a group of politicians whose ambitions for higher office were frustrated under the second-party system. Lincoln responds: "The Bible says somewhere that we are desperately selfish . . . I do not claim that I am any less so than the average of men, but I do claim that I am not more selfish than Judge Douglas" (see Johannsen, p. 314 [Lincoln]).

26. Again, the Southern fire-eaters as well as the most radical Northern abolitionists would dissent from this proposition.

27. Also, see Johannsen, p. 320 [Lincoln]. In this passage, Lincoln is, once more, responding to Douglas's charge that the "House Divided" doctrine constitutes an open invitation to civil war. Lincoln claims that once the options on the slavery issue are clearly stated the issue will be settled peaceably. This passage also suggests that all the Democrats who do believe slavery is wrong should really be in the Republican Party.

28. Lincoln is again quoting from his 1854 Peoria speech.

29. Although my analysis in this subsection will mirror George Forgie's controversial thesis about the central role of filiopiety in Lincoln's political thought, I will, in the end, discount that role. See George B. Forgie, *Patricide in the House Divided: A Psychological Interpretation of Lincoln and His Age* (New York: Norton, 1979), chap. 7. The overtly Freudian character of Forgie's thesis seems even more problematic to me.

30. Lincoln leaves open the question of why the founders felt they could not immediately change the *status quo* on slavery. He offers at least two explanations. (These explanations are not mutually exclusive.) One, they accepted the *status quo* because they otherwise could not have formed a constitutional union; two, they accepted the *status quo* because they had no legal authority to change it. Lincoln also assumes that the institution was foisted on the American colonies by the policies of the mother country. He cites Clay to this effect, but he

could just as well have cited Webster or any of a number of other American statesmen (see Johannsen, pp. 306–7 [Lincoln]; Webster X:65 ("Seventh of March Speech," 1850).

31. Douglas, however, would argue that even in Lincoln's own sense of the term the union was (is) divided on the slavery issue.

32. In effect, Lincoln throws the question back at Douglas. How could the union have survived so long if it had *not* been united on the slavery issue?

33. Also, see Johannsen, pp. 54, 132, 277–78 [Lincoln]. As additional evidence, Lincoln mentions the failure of the Constitution to explicitly refer to slaves. He was to make a more extended analysis of the founders' intentions toward slavery in his celebrated "Cooper Union" address. See Basler, pp. 517–26 ("Address at Cooper Institute, New York," 27 February 1860).

34. Not all historians are as charitable to the founders as Lincoln was on the question of whether they accomplished all they could have in putting slavery on the road to extinction. Cf. Freehling, "The Founding Fathers and Slavery," pp. 81–93; Edmund Morgan, "Slavery and Freedom," pp. 5–29.

35. For the most part, Lincoln and Douglas did agree on the constitutional status of slavery. They both supported the Southern states' exclusive authority over the institution within their own boundaries as well as a "fair" fugitive-slave law. Even on the question of slavery in the territories, Douglas advocated a noninterference policy more on the grounds of expediency than constitutionality (though by 1860 he clearly had shifted his position). This broad area of agreement on the Constitution was undoubtedly one of the reasons Lincoln pressed the dispute over the meaning of the Declaration of Independence to the forefront of the debates. See Fehrenbacher, *Prelude to Greatness,* pp. 107–12; Jaffa, *Crisis of the House Divided,* pp. 110–11; Zarefsky, *Lincoln, Douglas and Slavery,* pp. 127, 149.

36. Also, see Johannsen, pp. 219–20 [Lincoln]. Lincoln admits that some American statesmen, such as Calhoun, had previously denied the truth of the Declaration's claim that "all men are created equal." Cf. Calhoun IV:507–12 ("On the Oregon Bill," 27 June 1848). According to Lincoln, that position was at least more honest than the position that the claim only applied to white men. There seems little question that the founders intended the Declaration to refer to blacks in the sense Lincoln thinks they did. See Herbert J. Storing, "Slavery and the Moral Foundations of the American Republic," in *The Moral Foundations of the American Republic,* edited by Robert H. Horowitz (Charlottesville: University of Virginia Press), p. 217.

37. Also, see Johannsen, pp. 220, 306 [Lincoln].

38. While Forgie can explain Lincoln's need to go beyond the founders' principles and policies—indeed, that is the whole point!—he cannot explain how Lincoln did so, nor why he did so in such a different way than Douglas did. Furthermore, it seems to me that we can account for Lincoln's "need" more satisfactorily in terms of adjusting old principles to new circumstances than in terms of resolving some unresolved Oedipal tensions.

39. As one might expect, their lists of the previous crises of the union also differ. Douglas's list includes the War of 1812, the nullification crisis, and the current crisis; Lincoln's list includes the events surrounding the Mis-

souri Compromise (1819–20), the nullification crisis, the controversy over the annexation of Texas (1844–45), the events leading up to the Compromise of 1850, and the current crisis. I would naturally agree with Douglas that the nullification crisis was not caused by the slavery issue (although Lincoln concedes it was the one previous crisis that was not caused by efforts to spread the institution of slavery).

40. Again, the dispute partially concerns the relevant evidence. Most important, Lincoln's evidence includes the Missouri Compromise and Douglas's does not.

41. While this deterioration in public opinion is implicit to Lincoln's "House Divided" conspiracy theory, he explicitly refers to such a deterioration in his "Dred Scott" speech. See Basler, pp. 358–59 ("The Dred Scott Decision: Speech at Springfield, Illinois," 26 June 1857). (He does not quote this part of the speech during the debates.)

42. Here, I am using "racist" as the antithesis of "humanitarian." And I am definitely *not* saying that the racial attitudes of white Americans today are ideal, or even close to it.

43. Both Lincoln and Douglas were very sensitive to how their audiences differed by party as well as political geography. This sensitivity is the partial truth behind Douglas's "trimming" charge, a charge which, then, equally applies to himself. See Jaffa, *Crisis of the House Divided,* p. 368.

44. Lincoln is again quoting from his 1857 "Dred Scott" speech.

45. In subsequent debates, Douglas uses this and the previously quoted passage as evidence for his trimming charge (see Johannsen, pp. 195, 237, 264 [Douglas]). Some scholars see these passages as more faithful representations of Lincoln's own views and cite them as evidence of his racism. For example, see Fredrickson, "A Man but Not a Brother," pp. 46–47.

46. Obviously, this is the approach recommended by scholars who are more sympathetic to Lincoln's views and who argue that he was not as racist as he, at first sight, appears to contemporary audiences. For example, see Donald E. Fehrenbacher, "Only His Stepchildren," in *Lincoln in Text and Context: Collected Essays* (Stanford, CA: Stanford University Press, 1987), pp. 105–7. These scholars, thus, accept the gist of Douglas's trimming charge, though not its particulars nor its negative connotations.

47. See Thomas Jefferson, "Letter to John Holmes" (22 April 1820), in *The Portable Jefferson,* pp. 567–69.

48. In this case, Clay's influence is clear. Clay was a president of the American Colonization Society (although, again, Webster, like most politicians of his generation, at least verbally supported the society's efforts). Lincoln's 1852 eulogy of Clay largely focused on this aspect of Clay's life, and in it he quoted extensively from Clay's 1829 presidential address to the society. He also paraphrased part of that address during his debates with Douglas. See Basler, p. 276 ("Eulogy on Henry Clay Delivered in the State House at Springfield, Illinois," 6 July 1852); Johannsen, pp. 66–67 [Lincoln]. Cf. George M. Fredrickson, *The Black Image in the White Mind: The Debate on Afro-American Character and Destiny, 1817–1914* (Middletown, CT: Wesleyan University Press, 1971), pp. 6–21, 149–51.

49. Many Southerners argued the benefits of slavery for blacks in comparison to being members of such an underclass. Esp., see George Fitzhugh, *Cannibals All! Or, Slaves without Masters* (Cambridge, MA: Belknap, 1960).

50. For one thing, Lincoln, unlike Jefferson's letter, does not explicitly address the South's greatest fear—the fear that race wars would follow emancipation.

51. In truth, few abolitionists supported full social and political equality for blacks. See Litwack, *North of Slavery,* pp. 216–30. But, cf. Fredrickson, *The Black Image in the White Mind,* pp. 28–29. The crux of their differences with Lincoln was, then, the issue of immediate (or almost immediate) abolition. As compared to Lincoln, the abolitionists were clearly less sensitive to the constitutional and practical barriers to emancipation, the risks to the union, the question of compensation for the slaveholders, and the future circumstances of the freed slaves. See Fehrenbacher, "Only His Stepchildren," p. 108.

52. For the view that Lincoln was, virtually an abolitionist, see Dwight Lowell Dumond, *Antislavery Origins of the Civil War in the United States* (Ann Arbor: University of Michigan, 1939), pp. 106–14.

53. Also, see Johannsen, pp. 221, 282 [Lincoln].

54. Also, see Johannsen, pp. 225, 257, 319 [Lincoln].

55. See Stephen B. Oates, *With Malice Toward None: The Life of Abraham Lincoln* (New York: New American Library, 1977), p. 31; Johannsen, *Stephen A. Douglas,* p. 541.

56. Esp., see Diggins, *The Lost Soul of American Politics,* pp. 296–333; Greenstone, *The Lincoln Persuasion,* chap. 10.

57. The Jefferson quote is from his "Notes on the State of Virginia" (in *The Portable Jefferson,* p. 215). Lincoln uses it as part of his response to Douglas's charge that the founders would have been hypocrites to keep slaves while believing that the Declaration of Independence referred to blacks.

58. See Jaffa, *Crisis of the House Divided,* pp. 242–43.

59. Kloppenberg argues that American conceptions of civic virtue were overdetermined in this way. See Kloppenberg, "The Virtues of Liberalism," pp. 11–19.

60. Lincoln's frequent use of biblical allusions or quotes in his speeches is equivocal evidence of religious thought because of the way public discourse at the time so heavily leaned on the Bible as the one book with which most Americans were familiar. See Zarefsky, *Lincoln, Douglas and Slavery,* p. 18. The religious interpretation is much more persuasive in the case of the abolitionists, most of whom were (or at one time had been) members of one of the more radical Protestant sects. See David Donald, "Toward a Reconsideration of the Abolitionists," in *Lincoln Reconsidered,* pp. 24, 27, 29; Gilbert Hobbs Barnes, *The Antislavery Impulse, 1830–1844* (Gloucester, MA: Peter Smith, 1957).

61. Actually, this view of Douglas is an extrapolation. The republican revisionists themselves largely ignore Douglas and other political moderates, North and South, in order to explain the differences between the northern Republicans and the (more extreme) southern Democrats. These are the two political

forces they consider most relevant to explaining the outbreak of civil war. However, as I have already suggested, the real puzzle is not why the North and South divided but why the counsels of peace and moderation did not prevail within each section. Cf. Holt, *The Political Crisis of the 1850s,* p. 3.

62. Scholars who stress the religious elements in Lincoln's antislavery politics are especially unpersuasive when they portray him as a nonliberal. See Diggins, *The Lost Soul of American Politics,* pp. 298, 307, 325–26, 332. They peg the persuasiveness of their portrayal on the mistaken assumption that the kind of strong moral commitments Lincoln undoubtedly possessed could not have been rooted in liberalism. Jaffa seems to share that assumption, though the precise nature and sources of the nonliberal side of Lincoln's politics remain unclear in his account. See Jaffa, *Crisis of the House Divided,* chaps. 9–10, 14–15.

63. We should recall that for Douglas the rights of whites are not subject to the same calculus as the rights of blacks, although they presumably are subject to some calculus. A further complication is that for Douglas these calculations would take place on the state level and would almost certainly produce different results in different states. Lincoln, both because of his greater humanitarianism and because of his different reading of the exigencies of the union, would be more likely to perform just one calculation. In the following, I will ignore these complications to concentrate on their different understandings of liberalism.

64. When Lincoln recommended a constitutional amendment to Congress providing for gradual, compensated abolition, he felt compelled to rebut the argument that it would flood the North with freed slaves and depress the wages of white laborers. See Basler, pp. 685–86 ("Annual Message to Congress," 1 December 1862).

65. This priority is evidence of Douglas's republicanism. In this respect, but not in many others, he was more republican than Lincoln was.

66. As a matter of fact, this deprivation of citizenship rights could take place inside or outside of the institution of slavery; some white males and *all* white females experienced this deprivation in antebellum America. For opposing views on Lincoln's distinction between private and citizenship rights, cf. Fredrickson, "A Man but Not a Brother," pp. 52–53; Jaffa, *Crisis of the House Divided,* pp. 378–81.

67. See Gabor S. Boritt, "The Right to Rise," in *The Public and the Private Lincoln,* pp. 57–70. Lincoln's most striking statement of how slavery denied this right to rise was in a 1854 "Fragment on Slavery." See Basler, pp. 278–79 ("On Slavery," 1 July 1854?).

68. Douglas's position here was somewhat contradictory since, to his credit, he was unwilling to deny the humanity of the slaves. See Zarefsky, *Lincoln, Douglas and Slavery,* pp. 240–41.

69. See Basler, pp. 402–4 ("Speech in Reply to Douglas at Chicago, Illinois," 10 July 1858). This was Lincoln's reply to Douglas's initial campaign address, which had been given the previous day in Chicago.

70. Or, one might ask of the republican revisionists, how many Americans felt personally threatened by a Southern slavocracy?

71. During the Civil War, Lincoln not only saw the North's tremendous war effort as proof that Northerners had not lost their spirit of liberty, but he also

told them that they would ensure their own liberty by freeing the slaves. See Basler, pp. 620–21 ("Annual Message to Congress," 3 December 1861); p. 688 ("Annual Message to Congress," 1 December 1862).

72. The simplest (but also most racist) response to Lincoln's slippery-slope argument would have been to deny the humanity of the slaves; thus, in theory, totally isolating white Americans from the effects of slavery. Many Southerners made this response, even though Southern law never treated slaves merely as property. See Kenneth M. Stampp, *The Peculiar Institution: Slavery in the Ante-Bellum South* (New York: Knopf, 1963), chap. 5. Douglas, again to his credit, did not make this response. However, he did use the Declaration of Independence to isolate the unalienable rights of white Americans from the merely legal rights of black Americans, seemingly treating blacks as less human than whites.

73. Lincoln's idea of equal liberty came to fruition in the "Gettysburg Address." See Basler, p. 734 ("Address Delivered at the Dedication of the Cemetary at Gettysburg," 19 November 1863). Lincoln's understanding of liberalism clearly went beyond the founders' understanding, even if he, as was his usual practice, situated his understanding in their Declaration of Independence. This view departs from Jaffa's in that he sees Lincoln as transcending the founders' liberalism in *non*liberal directions. See Jaffa, *Crisis of the House Divided,* pp. 318–29.

74. The factors we considered in the previous section of this chapter—nationalism, pluralism, consensus, social realism, filiopiety, and history—also had some independent influence on Lincoln's position.

75. Lincoln's advocacy of gradual, compensated emancipation also respected interstate and interpersonal comity. As noted above, this constitutional consensus was fracturing exactly on the point the Republicans were most adamant: Congress's power over slavery in the territories.

76. At the Quincy debate, Lincoln observes that his party, while insisting on legally excluding slavery from all the new territories, does not insist on abolishing slavery in the District of Columbia, even though it feels the federal government plainly has the constitutional authority to do so. He, again, favors gradual, compensated emancipation (see Johannsen, pp. 254–55 [Lincoln]).

77. To reiterate, I do not mean to deny that the Lincoln-Douglas debates were, in part, a dispute over whether the slavery issue was (or should be) a question for national or local majorities.

78. Also, see Johannsen, pp. 225, 257 [Lincoln]. It is such statements which make Lincoln seem like a proponent of liberal or substantive democracy in distinction to Douglas's advocacy of majoritarian or procedural democracy. However, we must, in the first place, factor in Douglas's greater racism. He does not necessarily believe that the public liberty of white majorities takes precedence over the private liberty of white minorities. Both Lincoln and Douglas were liberal democrats with respect to white men. In the second place, we must take into account the possibility that their understandings of democracy primarily diverged in other ways than in terms of the distinction between substantive and procedural democracy. (The distinction is, in any case, a nebulous one. When is a political rule, such as popular sovereignty, "merely" procedural and when is

it also substantive?) Distinguishing between Lincoln and Douglas in terms of positive and negative liberty seems even less useful. Cf. Greenstone, "Political Culture and American Political Development," pp. 29, 42; McPherson, *Abraham Lincoln and the Second American Revolution,* pp. 61–62, 137–38.

79. Basler, p. 415 ("Speech in Reply to Douglas at Springfield," 17 July 1858).

80. Lincoln predicts (correctly) that Douglas will soon also be a sectional candidate according to his own criterion (see Johannsen, p. 223 [Lincoln]).

81. Douglas actually did not favor the decision. See Wells, *Stephen Douglas,* pp. 91–92, 112–13.

82. Brooks, of course, was Charles Sumner's attacker. Lincoln mentions Brooks in the context of his debates with Douglas because of Brooks's claim that the invention of the cotton gin had led to a fundamental change in attitudes toward slavery.

83. This analysis also calls into question the revisionists' reliance on the "House Divided" conspiracy theory, as was discussed in the introduction of this chapter.

84. Douglas, again, had other reasons for being silent. His belief that the preservation of the union required silence on the merits of the slavery issue was clearly an important factor. I, however, think his understanding of democracy was an equally important factor and one on which he and Lincoln fundamentally disagreed.

85. The "Freeport Doctrine" was, of course, Douglas's own adjustment to the *Dred Scott* decision.

86. Lincoln suggests the obvious way to try to overturn the decision: vote for Republican senators and presidents who will appoint justices unsympathetic to the decision (see Johannsen, p. 231 [Lincoln]). He also argues that Supreme Court decisions are only final for the parties to the case, not for other branches of government (see Johannsen, p. 255 [Lincoln]).

87. Also, see Johannsen, p. 55, 280–81 [Lincoln]. Lincoln reminds Douglas that two of his own political heroes—Presidents Jefferson and Jackson—did not accept the finality of previous Supreme Court decisions (see Johannsen, pp. 65, 232 [Lincoln]). Lincoln also insists that Douglas's "Freeport Doctrine" does, in effect, reject the *Dred Scott* decision as a political rule (see Johannsen, pp. 147–49, 280–81, 320–22 [Lincoln]).

88. Also, see Basler, pp. 300–301 ("Speech at Peoria, Illinois, in Reply to Senator Douglas," 16 October 1854). Consequently, Lincoln, unlike Douglas, rejected the "natural limits on slavery" thesis. See Jaffa, *Crisis of the House Divided,* pp. 117–21, 294–96.

89. Needless to say, Lincoln's pluralism is not the only reason that the Lincoln myth continues to resonate in American political culture.

90. Lincoln, in fact, was later to make an example of his commonness. Esp., see Basler, pp. 756–57 ("Address to the 166th Ohio Regiment," 22 August 1864). Although his historical argument is premature, Wood cogently discusses the political and intellectual consequences of the emergence of this new type of politician. See Gordon S. Wood, "The Democratization of the Mind in the

American Revolution," in *The Moral Foundations of the American Republic,* pp. 109–35. Wood expands on this argument in his new book, *The Radicalism of the American Revolution* (New York: Alfred A. Knopf, 1992).

91. For a different slant on the complexity of Lincoln's politics, see Ferguson, *Law & Letters,* p. 317.

Chapter Nine

1. The union even dropped out of political discourse. See Forgie, *Patricide in the House Divided,* p. 14.

2. See Howe, *The Political Culture of the American Whigs,* p. 302.

3. This discussion of the Progressive era is indebted to Bourke, "Pluralist Reading of James Madison's Tenth Federalist," pp. 278–85; Diggins, "Republicanism and Progressivism," pp. 572–98; Dorothy Ross, "The Liberal Tradition Revisited and the Republican Tradition Addressed," in *New Directions in American Intellectual History,* pp. 116–31.

4. See David B. Truman, *The Governmental Process: Political Interests and Public Opinion* (New York: Knopf, 1951), p. ix.

5. See Henry S. Kariel, *The Decline of American Pluralism* (Stanford, CA: Stanford University Press, 1961); Theodore Lowi, *The End of Liberalism: The Second Republic of the United States* (New York: Norton, 1969). (Of course, in my terms, Lowi's "the end of liberalism" is the end of pluralism.)

6. See Wilson Carey McWilliams, "Politics," *American Quarterly* 35 (1983): 19–38.

7. Cf. Michael Walzer, "The Communitarian Critique of Liberalism," *Political Theory* 18 (1990): 6–23.

8. See Richard F. Fenno, Jr., "If, As Ralph Nader Says, Congress Is 'The Broken Branch,' How Come We Love Our Congressmen So Much?" in *American Government: Readings and Cases,* edited by Peter Woll (Boston: Little, Brown, 1979), pp. 461–68.

9. See Barber, *Strong Democracy,* pp. 267–81; MacIntyre, *After Virtue,* p. 263; Sullivan, *Reconstructing Public Philosophy,* pp. 223–26. The republican revisionists, alternately, claim that republican themes still echo through American politics and that the nation's republican past is irretrievable. Cf. Ball and Pocock, *Conceptual Change and the Constitution,* p. 11; Lienesch, *New Order of the Ages,* pp. 207–14.

10. See Martha Derthick, "American Federalism: Madison's Middle Ground in the 1980s," *Public Administration Review* 47 (1987): 66–74.

11. Realizing that he faces a "Catch-22" situation in putting "strong democracy" into practice, Barber proposes instituting neighborhood assemblies in order to make local politics more meaningful. He hopes that political structures and public attitudes will, under the proper tutelage, evolve in the desired direction together. I am less hopeful. It, for instance, is not a passing fad that television has primarily been used to promote public "participation" in the form of consumer-related activities rather than, as Barber envisions, political ones. See Barber, *Strong Democracy,* pp. 273–78, 289–90.

12. See Ray M. Shortridge, "Voter Turnout in the Midwest, 1840–1872," *So-*

cial Science Quarterly 60 (1980): 617–29; "Nineteenth Century Turnout: A Rejoinder," *Social Science Quarterly* 62 (1981): 450–52. Shortridge argues that two factors principally account for the apparent long-term decline in the percentage of eligible voters who actually vote: (1) the extension of the franchise (hence, the importance, for purposes of historical comparison, of only sampling adult white males); and (2) the inaccuracy of nineteenth-century census data which seriously underestimated the population and, therefore, made turnout rates appear higher than they really were.

Index